T0222418

Anomaly!

Collider Physics and the Quest
for New Phenomena at Fermilab

Anomaly!

Collider Physics and the Quest for New Phenomena at Fermilab

Tommaso Dorigo

INFN, Italy

World Scientific

NEW JERSEY · LONDON · SINGAPORE · BEIJING · SHANGHAI · HONG KONG · TAIPEI · CHENNAI · TOKYO

Published by

World Scientific Publishing Europe Ltd.

57 Shelton Street, Covent Garden, London WC2H 9HE

Head office: 5 Toh Tuck Link, Singapore 596224

USA office: 27 Warren Street, Suite 401-402, Hackensack, NJ 07601

Library of Congress Cataloging-in-Publication Data
Names: Dorigo, Tommaso, 1966–   author.
Title: Anomaly! : collider physics and the quest for new phenomena at Fermilab /
    Tommaso Dorigo (INFN, Italy).
Other titles: Collider physics and the quest for new phenomena at Fermilab
Description: Singapore ; Hackensack, NJ : World Scientific Publishing Co.
    Pte. Ltd., [2016] | Includes index.
Identifiers: LCCN 2016015380| ISBN 9781786341105 (hc ; alk. paper) |
    ISBN 1786341107 (hc ; alk. paper) | ISBN 9781786341112 (pbk ; alk. paper) |
    ISBN 1786341115 (pbk ; alk. paper)
Subjects: LCSH: Colliders (Nuclear physics) | Particles (Nuclear physics)
Classification: LCC QC787.C59 D67 2016 | DDC 539.7/3--dc23
LC record available at https://lccn.loc.gov/2016015380

British Library Cataloguing-in-Publication Data
A catalogue record for this book is available from the British Library.

Desk Editors: Dipasri Sardar/Mary Simpson

Typeset by Stallion Press
Email: enquiries@stallionpress.com

Printed in Singapore

A mio padre

A Word from the Author

The standard model of electroweak interactions is one of the most striking successes of 20th-century science: a well-understood, coherent framework which allows physicists to calculate and predict the outcome of subnuclear reactions. Once combined with the theory of quantum chromodynamics, the model explains and describes in detail both electroweak and strong interactions, allowing us to understand the structure of matter at its deepest accessible level.

After an orgy of unexpected and surprising experimental results during the 1950s and the 1960s, the standard model established itself as a correct theory in the 1970s. Since then, particle physics has become much less chaotic. The contrast could not be starker, as the past 40 years have brought us an uninterrupted stream of pleasing, yet unexciting confirmations, with experiments finding results in more and more precise agreement with theoretical predictions.

In the last four decades experiments did occasionally detect anomalies, ephemeral effects, odd events, suggestive mass peaks, and other temporary deviations from theoretical predictions, but all those anomalous effects were eventually archived as statistical fluctuations, ill-understood systematic uncertainties, or *a posteriori* observations enhanced by the experimenters' unprincipled tampering with the data. This book tells the story of several of those anomalies, found by the CDF experiment during Run 0 (1988–1989) and Run 1 (1992–1996) of the Tevatron collider. CDF led the investigation of the high-energy frontier of particle physics from the late 1980s until 2010, when the Large Hadron Collider (LHC) experiments at the CERN Laboratories in Geneva made the Tevatron facilities obsolete.

CDF was run by a peculiar group of scientists, where two quite different ways of interpreting fundamental research clashed, keeping the experiment in a bipolar state throughout its lifetime. On one side was the conservativism of the Gatherers, happy with collecting measurements of standard model parameters and publishing strings of uncontroversial results. On the other side, the adventurous and open-minded attitude of the Hunters, who mined the data in new and creative ways in the hope to find new physics. The Hunters were especially attracted by the anomalies they found, and usually considered these as potential first hints of discoveries waiting to be claimed. Hunters and Gatherers managed to coexist only thanks to the continuous search for a scientific consensus by the spokespersons that alternated at the lead of the collaboration.

I participated in the CDF experiment since 1992 and followed closely the story of several anomalies unearthed by Hunters. This puts me in a good position to tell their story. However, care is required: by telling anecdotes involving the scientists who participated in the experiment, it is easy to be unfair to them or misrepresent their actions and motives. In addition, I owe to many of my former CDF colleagues a good part of my knowledge and insight in experimental particle physics. Whether they will acknowledge that the stories contained in this book really had to be told, one way or the other, remains to be seen. A special mention needs to be made here of Paolo Giromini, who is one of the main characters of this play, and is — or was, if he takes this book the wrong way — a good friend of mine. As Oscar Wilde once wrote, "Every great man nowadays has his disciples, and it is always Judas who writes his biography." I have certainly been a disciple of Giromini in some way. In this book, I have made an attempt at not being a Judas to him, but if Wilde is correct I must have failed somewhere.

I believe that some instructions are needed to make the best use of this book. Learning even only the rudiments of particle physics is a highly demanding intellectual activity. In order to enable you to quickly get personal with cutting-edge science, I have attempted to be as inaccurate as my conscience allows me; accuracy is the enemy of comprehension when complex matters are discussed. Still, occasionally you will find paragraphs heavy with explanations you might not need. If you run out of patience, jump to the following paragraph and you will likely find out that things make sense again. Words or expressions you are unlikely to have ever met before are written in *italics* the first time they are used. A "subject index" at the end of the book helps locating the definition of those terms.

A word needs to be spent also on the accuracy of the stories and anecdotes reported in this book. Although in all cases I have tried to collect as much information as possible by interviewing the involved parties, it is very likely that some of my reports are biased, incomplete, or partly incorrect. This is okay, I think, as their goal is to teach some physics in an entertaining way rather than to contribute to the history of science. For this reason, no bibliographical references are given. The sources of most of the facts discussed in this book are conversations I had with over a hundred of my former CDF colleagues.

The book is divided into three parts. The first includes Chapters 1–3. Chapter 1 is meant to provide the reader with some basic notion about particles and forces, the currently accepted theory of fundamental physics, and a glimpse at its possible extensions; Chapter 2 discusses the experimental apparatus on which the story is based; and Chapter 3 focuses on a few early stories which I decided to keep separated from the others, as they integrate better with the material of the previous chapters. The second part of the book includes Chapters 4–7, which offer a history of the tortuous path followed by the experiment to discover the top quark. Finally, Chapters 8–12 are dedicated to the description of anomalous effects found in the data collected by the CDF experiment in the 1990s, and the controversies they generated.

I believe this book can be read back to back by anybody who is patient enough to start from square one and work his or her way through the introductory material provided in the first few chapters, before plunging into the second and third parts. However, readers already familiar with high-energy physics might consider skipping Chapters 1 and 2 altogether.

Tommaso Dorigo, 2016

Prologue

On a fine afternoon in the early summer of 1998 you are driving on Highway 88 from downtown Chicago to the Fermi National Accelerator Laboratory. You exit on Route 59 heading North, soon make a left on Batavia road, and enter the laboratory from the East entrance. As you proceed West, you notice the remarkable skyline of Robert Rathbun Wilson Hall, planted in the middle of a pleasant country scenery of lakes, woods, wildlife, and the occasional piece of junk metal — at least, that is what the parts of past experiments and accelerators placed here and there might look to your untrained eyes, if you are not a particle physicist. Wilson Hall, a 16-story building made of concrete and glass commonly referred to as "the Hirise," serves as the administrative center of the lab and hosts offices and facilities for the Fermilab personnel, such as meeting rooms, a cafeteria, a library, and a visitor center.

Proceeding toward the Hirise, you soon find on your left a large industrial building, painted in bright orange. That building hosts the CDF detector, along with the infrastructure needed to operate and service it, and the control room of the experiment. Next to the detector building stand the CDF portakamps, a set of office trailers connected together in a comb-like structure. Dirty white in color and shabby-looking inside out, the trailers are the true headquarters of the experiment. As you enter through one of the many access doors, you get to walk through narrow, poorly lit corridors, lined with posters of past conferences, letter-format notices left hanging way past their expiration date, and paper clips of all kinds, from science-themed cartoons by Gary Larson to fancy pictures of distant galaxies. Each trailer is divided into several 60-square-feet offices, which the resident physicists stuff with their own junk. Workstations sit next to rollerblades; piles of papers crowd every corner; broken coffee makers rest unused on dirty shelves since God-knows-when.

A no-smoking policy is enforced site-wide, but the trailer owned by the Frascati group is for all practical purposes a free zone; despite the clear security issue, no building manager has ever won that battle. Indeed, walking to the end of the corridor toward the office of the group leader, Paolo Giromini, you are likely to plunge into a standing haze of bluish smoke, generated and replenished daily by a pack of Winston cigarettes. The office behind the door is wider than most of the others, yet a decade-long stratification of papers, cigarette butts, tea cups, books, logbooks, and scribblings on yellow paper has conquered the whole length of the red L-shaped desk which runs along the wall on two sides of the room; above it, two full rows of white plastic shelves precariously clinging to the plywood walls suffered a similar fate. On the chair in front of a bulky monitor sits Paolo Giromini, a research director of the Italian National Institute for Nuclear Physics (INFN), and the *de facto* head of the Frascati physicists in CDF. He wears thick square glasses; his brown hair is straight and combed on one side. Paolo wears his usual work suit: blue jeans one size too wide, a green Lacoste T-shirt, and a cigarette between his fingers. He is staring at the large screen with a perplexed look.

Giromini is a brilliant, indefatigable, and stubborn scientist. At over 50 years of age he still has the stamina to write analysis code 12 hours a day, like only a graduate student or a young post-doc would find normal to do. But he is also one of the few colleagues around who has the know-how to look for mistakes in theoretical calculations — and he does occasionally find some. Such a skill is not at all common in experimental physicists. You might expect that experimentalists have studied quantum field theory and theoretical physics during their undergraduate studies, so they should be capable of computing cross sections of particle reactions and decay rates. Yet, those skills are very hard to preserve as years are spent writing software or building detectors.

Giromini would have the skills and the experience to be a true leader in the experiment, but he very much prefers to spend his time doing physics rather than dealing with bureaucracy and paperwork. Besides, he also has his own, quite human, limits: in particular, his ego is large. That is a quite common trait of particle physicists. But when large egos have to get past each other or walk side by side along the narrow corridors of the portakamps, some friction between them is to be expected. And friction, between Giromini and the rest of CDF, is the rule. He considers at best mediocre most of his collaborators, and he makes no bloody mystery of it. After all, he comes from Carrara (a small town in Tuscany), and Italians of Tuscan origin are typically blunt beyond repair. Alas, bluntness is a sure recipe for trouble within a large scientific collaboration of 50 among the most prestigious universities and research institutes in the world.

It is a hot afternoon in Batavia, and the air conditioning fails to remove the heat generated by computers. Giromini is not sweating: it looks like even his skin glands are taking part to the deep thoughts that trapped him in, while a fan blows smoky air against his back. The nicotine level in his blood is dangerously low; he has only few cigarettes left in a battered soft pack, and he cannot find his matches. He swears to himself, shuffles papers around, finally finds the emergency lighter on the floor under the table, and nervously lights up the umpteenth Winston of the day. Yet, the puffs of smoke do not clarify matters. The screen of his desktop computer is crammed with a bunch of graphs that would mean little to anybody else. Those graphs describe something very weird about the events he has isolated.

To a trained physicist, any oddity or anomaly is immediately worth full, undivided attention: the challenge is on, the features of the data require an explanation. His or her reaction to a surprising observation is not really to make a phone call to *Nature* (the magazine, not the bitch). The countless tentative discoveries made in the past, all later turned out to be artifacts or fluctuations, taught him or her the lesson. What needs to be done is to go back and check how the observed feature was computed, how the data were collected, and everything in between. Once, twice, three times, swearing silently, until the cause emerges. The analysis needs to be checked step by step; each line of code needs to be scrutinized with the magnifying lens; assumptions have to be questioned; approximations have to be eliminated. It is usually a painful, repetitive process, and a humbling one.

More often than not, it will yield the discovery of a bug in the code; in the remaining cases, it will reveal that what was shown in the histogram was a perfectly normal feature of the data, and the alleged weirdness was only due to lack of insight of the observer.

Insight, however, is not an issue in this case. Giromini understands well the physics of the studied events. Nor does he even consider checking the code again. He knows everything is alright: Fotios Ptohos and his other collaborators have done all the dirty work. The young researchers working under Paolo's supervision have filtered, reconstructed, corrected, retuned, and checked in every way a well-known sample of data. All that the group leader is doing is applying some apparently innocuous selection cut to the original sample, in order to focus on a particular subset of the events. And those 13 events he rounded up possess quite startling properties. They are a true *lusus naturae*. Maybe nature, the bitch, does have a sense of humor.

In front of Giromini stands the distilled result of three years of work spent in the attempt of understanding the features of data used for measurements of the properties of the top quark. The result reveals a tantalizing anomaly and opens a door to unexplored territory. But, in science, the opening of one door usually reveals others, closed by more complicated locks. Those three years have materialized a puzzle which is going to keep him busy for a lot longer. Will his stubbornness be sufficient for the arduous task? The Italian physicist does not yet nearly imagine the amount of trouble he has come to face. For now, he is overheating his brain in search for an explanation, a new physics model capable of explaining the features of those 13 weird events.

Contents

Chapter 1

The Standard Model and Beyond

A famous quote attributed to Albert Einstein says "The most incomprehensible thing about the world is that it is comprehensible." If one looks at the depth of our current understanding of subnuclear physics, one has to agree with the great Einstein. Humans have succeeded in explaining natural phenomena down to mind-bogglingly small distance scales, and such is an extraordinary accomplishment; yet, even more surprising is the fact that a theoretical explanation was there to be unearthed in the first place. The *standard model* is that theory. It results from unifying together into a single elegant construction the mathematical descriptions of phenomena involving three apparently unrelated natural forces: the *electromagnetic interaction*, which holds atoms together and describes the forces acting on electrically charged bodies; the *weak interaction*, which helps make stars burn and causes the radioactive decay of heavy elements; and the *strong interaction*, to which we owe the stability of atomic nuclei. These forces act on matter particles, so it is useful to introduce the latter before we explain how forces work.

Quarks and Leptons

What is matter made of? This question has fascinated scientists for thousands of years. To answer it, we need to search for matter's most elementary constituents. In the fourth-century BC, the Greek philosopher Democritus called these bodies ἄτομα (atoms: "that cannot be cut"). Today, we identify the elementary constituents with a short list of point-like subatomic particles. As far as we know, Democritus was right — it is us who fallaciously identified his atoms with what turned out to be complex composite systems!

Subatomic particles come in two very different varieties, depending on an intrinsic property they possess called *spin*. Mathematically spin is a *vector*, i.e., a quantity oriented along some direction in space. Spin behaves like angular momentum, the vector used in classical mechanics to describe the rotation of a body around some axis. Yet, for elementary particles, which follow the rules of quantum mechanics, spin is not to be mistaken for the rotation of the body around itself. Spin in quantum mechanics is measured in units of *h-bar*, the so-called "reduced Planck constant." If one counts spin in h-bar units, one observes a very different behavior in particles possessing integer and half-integer spin. Particles with integer spin are called *bosons*. The rules of quantum mechanics show that the presence of a boson in a point of space makes it more probable to find in its whereabouts additional identical bosons. In contrast, particles with a half-integer value of spin, called *fermions*, behave the opposite way: the presence of a fermion in a point of space mathematically excludes the possibility to find there a second one with identical properties. The solitude of fermions is one of the most important ingredients of atomic physics: electrons, the fermions we are most familiar with, can occupy the same orbit around a nucleus at most in pairs, when only their oppositely aligned spins make them distinguishable. This principle, first recognized by Wolfgang Pauli, forces electrons to occupy larger and larger orbits pairwise; as a result, atoms with a different number of electrons display different chemical properties. Without Pauli's principle, all electrons would settle in the lowest energy orbit around atomic nuclei; that way, chemistry would be trivial and life would not exist!

Matter is made of fermions, so we need to study those particles in detail in order to understand nature. Elementary fermions are organized in two families: *quarks* and *leptons*. Quarks were hypothesized independently by Murray Gell-Mann and George Zweig in 1964 in order to explain with few simple constituents the scores of subatomic particles observed by particle accelerator experiments. Each discovered particle could be described as a combination of three quarks or as a quark–*antiquark* pair.

All one had to allow for was the existence of three different kinds of quarks, with different physical properties, plus *antimatter* copies of each, endowed with opposite properties (the "anti-" suffix identifies them in the following).

The three quarks were called *up, down,* and *strange*. They were posited to carry fractional electric charges in electron charge units. The up quark had a positive charge of +2/3, and the other two a charge of −1/3. The proton could then be described as the combination of two up quarks and one down quark; the neutron instead could be made by an up quark and two down quarks. As for e.g. the positively charged *kaon*, a light unstable particle produced in copious amounts by the first accelerators, its properties matched those of an up-antistrange pair, a quark–antiquark combination. The above description enabled the fitting of observed particles displaying similar properties into tidy sets called *multiplets*. That cataloguing scheme was a radical simplification of the subatomic world, but initially few believed in the physical reality of quarks, as there was no direct proof of their material existence. They were considered no more than a useful mathematical description.

Quarks were accepted as true physical objects only after the November 1974 discovery of the J/ψ, a new particle simultaneously seen at the Stanford and Brookhaven laboratories. It is a fun fact that the confirmation of the three-quark model came by discovering … the fourth quark! In fact, it had been suggested by theorists already in 1970 that the three-quark model needed an upgrade to four, or even six quarks, in order to explain some apparent inconsistency in the observed properties of the known particles. In particular, a fourth quark called "charm" had to exist and weigh roughly one-and-a-half proton masses. The J/ψ, weighing just a bit over three proton masses, and exhibiting the exact properties that a bound state of two heavy quarks had to possess, was immediately recognized as a charm–anticharm quark system. The discovery wiped overnight all the doubts about the correctness of the quark model.

Nowadays, we recognize quarks as the basic constituents of protons and neutrons as well as of all other *hadrons*, most of which are unstable particles with funny names. Hadron comes from the Greek word αδρὸς, meaning "dense, strong": all particles made of quarks are called that way as they experience strong interactions. Quarks come in six *flavors*: six different kinds, each with its own characteristics, like assorted candies in a box. They may be fruitfully divided into three families, *doublets* containing two quarks each. The first doublet includes the up and the down, which as I mentioned above are all what is needed to make protons and neutrons. Yet there are two more families, made of heavier quarks which come into play only in phenomena where a large amount of energy is released. The second family includes *charm*, which is a heavy replica of the up quark, and *strange*, the down-type partner of the charm quark. Still heavier are the members of the third family: the up-type *top* and down-type *bottom* quark. In unambiguous contexts, quarks are labeled by their initials: *u, d, c, s, t*, and *b*.

To complete the above terminology, let me add that three-quark combinations are called *baryons* (from the Greek word $\beta\alpha\rho\dot{\upsilon}\varsigma$, heavy). While not elementary but composite, baryons are themselves fermions, as they possess a half-integer spin like their constituents. This is because the three spins of the quarks follow the rules of addition of quantum angular momenta, which state that combining three 1/2 spins can only give a total of 1/2 or 3/2. To understand why, imagine you are on a ladder, whose steps are half a foot tall. If you are asked to make three random steps, you can either change your elevation by three-halves of a foot, going all the way up or all the way down; or by one-half, e.g. by going up, then down, and then up again. Quark–antiquark combinations like the kaon K and the *pion* π are instead called *mesons*. The word meson also comes from Greek — in this case $\mu\dot{\varepsilon}\sigma o\varsigma$, which means middle: the mass of the lightest mesons, the first ones to be discovered, was found to lay in-between the mass of the electron and that of the proton. Mesons are composite bosons as they have integer spin: the combination of two 1/2 spins can only give 0 or 1, as you can immediately prove with the ladder example mentioned above.

Leptons are the other half of the story. The word "lepton" also has a Greek etymology: leptos ($\lambda\varepsilon\pi\tau\dot{o}\varsigma$) means "thin." Within each generation, leptons are lighter than quarks. In full similarity to quarks, leptons also make three families of two elements each; the first family includes the *electron* e and the *electron neutrino*, ν_e, the second counts the *muon* μ and the *muon neutrino*, ν_μ, and the third has the *tauon* τ and the *tauon neutrino*, ν_τ. Muon and tauon are heavier replicas of the electron and are equally charged. The electrically neutral neutrinos instead have masses so small that it has not been possible to measure them yet. Unlike quarks, leptons

Generations

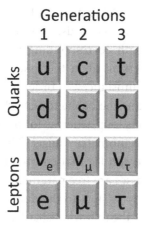

Figure 1.   The three families of matter: quarks (the first two rows) and leptons (the bottom two) are organized in a scheme of three generations (the three columns of four elements labeled "1," "2," and "3"). Note that for each of the 12 particles listed in this scheme, one has to also consider the corresponding antiparticle, of equal mass and spin and opposite electric charge. In addition, each quark comes in three different color charges.

do not aggregate into stable bound states because there is no glue to keep them together: they do not feel the strong force. The electron, muon, and tauon only experience weak and electromagnetic interactions. Neutrinos do not even feel the electromagnetic interaction and obey only to the weak force. The three families fit tidily in a three-column grid.

The Enigma of Fermion Masses

The regularity of the scheme shown in Figure 1, which collects *all* known elementary fermions in three families, is at odds with the hugely varied masses that fermions exhibit. Mass is a mysterious intrinsic property of matter corpuscles. We are accustomed to confuse the mass of an object with its weight, since the two quantities are proportional to each other as long as we remain in the roughly constant gravitational field of Earth's surface. Yet, mass is not a force: rather, it is the attribute of a material body specifying how hard it is to change its status of motion. The larger the mass of an object, the stronger the pull one has to apply to set it in motion with a given acceleration, or to modify its trajectory.

To visualize how different the masses of fermions are, try to imagine measuring fermion masses with a yardstick as if they were lengths in space. Let us agree that the length corresponding to the mass of the lightest charged fermion, the electron, fit in the first notch of our scale — say one inch. Under such circumstances, in order to measure distances corresponding to masses like the one of the top quark, we would need a yardstick about eight kilometers long, whereas for neutrinos we would need to discern lengths in the nanometer range! The standard model gives neither any clue on the reason for that enormous range of values nor for their pattern: fermion masses are free parameters of unknown origin.

It would be rather silly to stick to the dubious analogy of measuring masses with a yardstick. In the following chapters, you will instead become familiar with a more standard treatment of mass and energy, two quantities which can be sized up with the same units. All works in multiples of the *electron-Volt*, the energy imparted to a particle carrying the electric charge of one electron by moving it "upstream" through a potential difference of 1 V. The electron-Volt (usually written eV) is a quite small amount of energy: its order of magnitude is the energy of atomic transitions between nearby energy levels. In this book we will often use a multiple of the eV called GeV, where the G stands for giga-, the prefix indicating a billion. The proton mass is close to a GeV, and so are the masses of many other composite bodies whose study has opened our way to the understanding of subnuclear physics.

Now armed with a handier measurement unit, let us give a second look at the mass hierarchy of the fundamental objects we have encountered so far. Neutrinos are the lightest fermions, and all we know is that they are much lighter than 1 eV. Next up is the electron, with a mass of 511,000 eV, i.e., 511 keV, where the prefix k stands for a thousand. The muon is still 200 times heavier, at 105 MeV, where M stands for a million. The tauon, the heaviest lepton, is some further 17 times heavier at 1.77 GeV. As for quarks, the lightest is the up quark, which weighs about 3 MeV, and the heaviest is the top quark, with a mass of 173 GeV. In-between lay the down quark (7 MeV), the strange quark (90 MeV), the charm quark (1.2 GeV), and the bottom quark (4.2 GeV). Those given above are approximate numbers; in particular, the masses of all quarks except the top cannot be measured directly, as they can only be studied within hadrons, the bound states quarks form by combining in pairs or triplets. Like a cat that would not be convinced to stay still on a scale, quarks force us to use indirect methods to infer their masses. You can estimate your cat's mass by subtraction, measuring your own weight with and without the cat in your arms; similarly, you can size up quark masses by comparing the masses of similar hadrons that contain them or not.

Vector Bosons: The Interaction Carriers

We have learned that matter at a fundamental level exists in several different varieties: different kinds of fermions. They truly are the stuff that matter is made of; however, fermions could not be held together without the help of some interaction binding them together. The forces that organize leptons and quarks in stable structures are carried by bosons, the particles of integer spin introduced at the beginning of this chapter. The three fundamental interactions described in the standard model — electromagnetic, weak, and strong — take place via the exchange of three classes of vector bosons: respectively, *photons*, *W* and *Z bosons*, and *gluons*. The suffix "vector" clarifies their function of mediators of the forces, but it is actually meant to describe the unit of spin that these particles possess.

Electromagnetic interactions found a detailed theoretical description within the framework of *relativistic quantum mechanics* (the combination of Einstein's theory of special relativity with the theory of quanta), thanks to the work of many brilliant minds during the first half of the 20th century. That effort culminated with the contribution of Richard Feynman and Julian Schwinger in 1957, which earned them the 1965 Nobel Prize in Physics. The theory of the electromagnetic field, which was called *quantum electrodynamics* (QED), was thus the first one to be fully "comprehended" in the sense that Einstein would give to the word. Electrically charged bodies feel each other by exchanging photons, which are the "field quanta" transmitting the interaction. Photons are massless neutral particles. Depending on their energy, they make up radio signals, visible light, or X-rays.

Weak interactions exist in two kinds, called *charged currents* and *neutral currents*, respectively, highlighting the fact that weak processes may or may not change the electric charge of the particle on which they act. Charged currents are mediated by the exchange of W bosons, whereas neutral currents are mediated by the exchange of Z bosons. Charged currents were first formulated in the 1930s by Enrico Fermi to explain the *beta decay* process whereby the neutron turns into a proton, an electron, and an electron antineutrino. In contrast, neutral currents were only formulated 30 years later.

Whether neutral currents existed was a good question waiting to be asked: indeed, science progresses by asking critical questions rather than answering inessential ones. None of the hundreds of particle reactions and decays studied from the 1940s to the early 1970s manifested effects which called for the exchange of a neutral vector boson. That is precisely why the prediction of the existence of neutral currents, put forth by Sheldon Glashow, Abdus Salam, and Steven Weinberg in the 1960s, was a gigantic intellectual achievement. With the introduction of the Z boson, electromagnetic and weak interactions could be described together in a single powerful expression that summarizes what we today call the standard model: the theory thus obtained a formal unification of electromagnetic and weak forces. With the 1973 discovery of weak neutral currents, the theory was accepted as correct, and in the following years it received countless experimental confirmations.

A theoretical description of strong interactions could be added to the model after quarks were recognized as real entities. The strong force was understood to be the glue keeping nuclear matter together. No formal unification between strong and electroweak forces is, however, contained in the standard model: the key to a unification of strong and electroweak forces is believed to be hidden in processes that take place at enormously high energies, ones unlikely to be ever probed directly at man-made accelerators. Despite this, the standard model enables a synthetic description of elementary particles and their full set of strong as well as electroweak interactions. The way this is made possible is by describing fermions as *irreducible representations* of the product of three *symmetry groups*, corresponding to the three interactions. This means that each particle is uniquely determined by specifying its symmetry properties under a set of transformation laws; such a set is called a group when it possesses certain mathematical properties.

We can use a very simple example to clarify the above statement. Do you look the same in a mirror? If you do, you possess apparent left–right symmetry. In that case, your image is said to be *invariant* under the set of transformations composed by the two operations "left→right" and "right→ left": it does not change when transformed by the mirror. Simpler objects may possess higher levels of symmetry. A sphere looks the same regardless of how you turn it and by how much: the group of all conceivable rotations in three dimensions is a richer mathematical set than the one containing just left–right inversions, but the idea is the same. If I come to you with a bundle in my pocket and say it is an object which looks exactly the same from every direction, you immediately guess it is a sphere! That shows how one can describe the features of objects using their transformation properties: the latter capture the essence of the object's shape.

One may thus think of a symmetry group as the collection of transformations one can make on an object; this may include changes in its properties. A particle can be translated in space by a given length, or rotated by a given angle around some axis: these are conventional moves in three dimensions, affecting extrinsic properties such as position or orientation. But a particle can also withstand less intuitive transformations, performed in the space described by its intrinsic properties. For instance, electric charge is an internal attribute which cannot be changed by rotations or translations. One may, however, change it in the standard model using the transforming power of a charged-current weak interaction. By interacting with a W boson, quarks and leptons turn into their partners in the doublet, which have a different electric charge. That is precisely what happens in the aforementioned process of beta decay, when the down quark contained in the neutron transmutes into an up quark by emitting a W boson (the W then materializes the electron and the electron antineutrino). In summary, by saying that a particle is specified by its transformation properties under the symmetry groups of strong, weak, and electromagnetic interactions, we mean that the particle possesses a small set of attributes which define its phenomenology: they determine the way it behaves under the action of those three forces. The forces are sensitive to the attributes that the three groups describe: these are called *color charge, weak isospin*, and electric charge, respectively.

Similar to QED, the theory of strong interactions is called *quantum chromodynamics* (QCD). As far as QCD is concerned, quarks of different flavor are identical; the only feature of quarks that QCD forces are sensitive to is that they come in three "colors." You may call these red, green, and blue, provided you acknowledge that it is just a convention. The interaction of colored objects is mediated by eight *gluons*. Like the photon, gluons are electrically neutral, massless vector bosons; however, the similarities between gluons and the mediator of electromagnetic forces stop there. In fact, gluons are bi-colored, so they themselves carry a color charge like the particles subjected to their interaction. The color charge of a gluon can be red–antiblue, green–antired, and so on. Hence, gluons also act as sources of the force field and interact with one another as quarks do, i.e., by exchanging other gluons. This feature makes the phenomenology of QCD quite different from the one of QED; because of it, strong interactions become stronger as the distance between two color charges increases. This makes it impossible for a quark to exist as a free entity: quarks may only exist within colorless hadrons.

To obtain a colorless object from colored quarks, one way is to put together a red, a blue, and a green one: the three colors blend together in a colorless object and nullify each other just as do gin, Martini, and lemon juice in a carefully prepared drink. But handling quarks might prove harder than you thought. You could imagine picking up a quark from inside a proton and pulling it out. The attempt to separate the quark from its partners works in close similarity to the extension of a spring: you need to keep pulling as you extend the "color string" further to get the quark out. At the expense of your effort, the potential energy of the system increases with the quark distance from its partners in the proton, so you cannot go very far with it: you would have to supply an infinite amount of energy to walk away with your quark. But, in addition, there is a further problem.

A very general law of nature dictates that physical systems spontaneously seek the configuration of minimum potential energy. So as you pull the quark far out enough, the materialization of an extra quark–antiquark pair becomes energetically favorable with respect to a further extension of the original string. The two extra particles "pop into existence" from the vacuum and they break the lines of the color string into two. The new quark is sucked inside the proton and takes the place of the removed one, making the color charge budget of the proton null again; and the new antiquark binds with the original quark you were pulling out. What is left in your hand is a colorless meson. To create it, you had to supply at least an energy corresponding to its mass. So here it is — the second way to obtain colorless objects! Of course, to pull a quark off a proton, we do not use some fancy brand of precision pliers; rather, we rely on an energetic collision. In that case, the kinetic energy imparted by the collision to the quark kicked off the proton is the source of the energy required to materialize new particles.

Hidden Symmetry and the Higgs Boson

Just as you started to think you understood that the standard model fits all known elementary particles in a beautiful scheme ruled by symmetry — a symmetry of strong interactions acting equally on quarks of different flavor, a symmetry of electroweak interactions organizing quarks and leptons in three families, and also a symmetry between charged and neutral electroweak interactions — we need to deconstruct that concept a little. When we say that weak vector bosons are *unified* with the photon, we mean that they fit in a common mathematical description: a group of symmetry, as we have seen. In spite of that, this "electroweak unification" is not apparent in nature, since those particles are quite different in one important way: the photon has zero mass, whereas W and Z bosons are very massive.

The large mass of weak bosons is the reason why weak forces are called that way: the mass of a mediator determines its interaction range and strength. To understand how a massive mediator may be less effective and act more weakly than a massless one, compare a cup of hot chocolate to a raw chocolate bar. The hot vapor of the former disperses around many very small, light-weight particles, which you can easily smell from a distance; the latter can only release a few small specks of solid chocolate if you get close and inhale powerfully. The specks are much more massive and less copious than the molecules evaporating from the cup and are thus incapable of carrying the chocolate interaction far away; furthermore, the chocolate smell you may experience even at small distance from the bar is less intense, because of the small rate at which the bar releases a speck of chocolate when you sniff it!

The properties of the smell of liquid and solid chocolate may be likened to the properties of electromagnetic and weak interactions *at low energy*, where electromagnetic interactions appear more intense than weak ones. Now, however, let us imagine, for the sake of arguing, that we construct a computerized sniffer that analyzes the odor of solid as well as liquid materials. The sniffer works by taking the material under test, vaporizing it, and analyzing the composition of the produced vapor. Such a device will find that cup and bar of chocolate produce the same intensity of chocolate smell; that is to be expected, in fact, since the molecules of the two samples are the same. Likewise, electromagnetic and weak interactions acquire similar strength *at very high energy*, once the different masses of chocolate particles and solid specks — pardon, of photons and W/Z bosons — become irrelevant.

Weak interactions are indeed weak at low energy, as they are mediated by massive particles unable to propagate through long distances; thus, electromagnetic and weak interactions display a different phenomenology. It turns out that the large mass of W and Z bosons is a crucial feature of the standard model. If those particles were massless, the symmetry of electromagnetic and weak interactions would be more manifest, but the theory would contain a destructive inconsistency: it would predict an infinite reaction rate for processes involving the interaction of W and Z bosons at high energy. An electroweak theory with massless vector bosons is said to be *non-renormalizable*: it does not lend itself to the prediction of observable physical quantities, as the calculations fail to converge to finite results. The blessing for the electroweak unified theory came in 1971, when the Dutch theoretician Gerard 't Hooft proved that indeed, with massive vector bosons, the model is renormalizable. Until then, the 1967 paper by Salam and Weinberg had received little or no consideration by the theoretical physics community.

Renormalizability is an important feature of the standard model, but it is only granted if W and Z bosons are massive. On the other hand, the symmetry properties of the theory, which allow the elegant unification of electromagnetic and weak forces, appear to require all bosons to be massless. The solution to this riddle was discovered already in the 1960s, with the monumental intuition of a few theorists that a *hidden symmetry* mechanism was at work. A symmetry of a physical system is said to be hidden when the configuration of minimum energy of the system — its *fundamental state* — fails to enjoy the symmetry properties of the theory.

Imagine that you place a thin cylindrical plastic stick vertically over a flat surface and hold it there by applying a small downward pressure on its

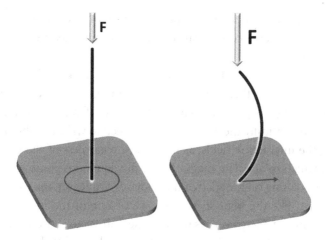

Figure 2.   Left: A plastic stick can be held vertically on a table by exerting a force *F* (downward arrow) on its tip; an equal reaction opposite to it (not shown) is exerted by the table. Right: if the force *F* exceeds a critical value, the state of lowest energy of the system becomes one where the stick is bent along one of an infinity of possible directions on the horizontal plane. The forces are still symmetric, but the system is no longer so.

top tip (see Figure 2): the pressure causes friction at the points of contact of the stick with finger and table, and the stick remains perfectly vertical, oblivious of anything but its own axis of symmetry. The system is symmetric around the vertical axis, in the sense that no part of it allows you to distinguish the different horizontal directions: the force that you exert on the stick and the force between stick and table both look the same regardless of where you sit around the table. If, however, you apply a larger pressure on the tip, this causes the unbent vertical configuration of the stick to become one of unstable equilibrium. An arbitrarily small perturbation of the system eventually causes the stick to bend and make an arc, whose elastic tension counters the extra pressure. The arc is now extended along one among the infinite possible directions on the horizontal plane. The new state of equilibrium is energetically favorable with respect to the unstable symmetric situation, but in exchange for the lower energy the system has had to give up its symmetry properties. The force applied by the finger and the opposite reaction of the table are still symmetric, as they require no mention of the horizontal coordinates to be described. Yet, the symmetry of the system is *hidden* by the particular choice that the stick has made in bending along one particular direction: a choice of one among an infinite set of minimum energy configurations, only distinguished by the direction of elongation.

The same mechanism described for the plastic stick applies to the fundamental state of the theory of electroweak interactions, i.e., its vacuum, a void containing no particles. That is the state of minimum energy. To describe mathematically the vacuum, we still need to write out the equations describing the interaction fields, as those exist regardless of the presence of matter on which to act. The magic is that the equations exhibit symmetry, like the forces acting on the stick of the above example; yet, the vacuum state is asymmetric, like the bent stick. A choice of one of the infinite possible states of minimum energy makes three of the four vector bosons of the electroweak theory — the positively and negatively charged W bosons and the Z — become massive, whereas the photon remains massless!

The price to pay for the miracle of a symmetric theory with an asymmetric vacuum state is the introduction of one additional ingredient in the mathematical formulation of the theory: the Higgs boson field. The Higgs boson is the particle that gives the electroweak formulas their "elasticity," the freedom to bend in one particular direction. Once the Higgs field is added to the mathematical equations, the potential energy of the symmetrical position becomes larger than that of an infinity of equivalent, stable configurations analogous to those of the bent stick. The stable configurations represent equivalent possible choices for the true vacuum of the theory, since they correspond to the minimum energy. But only one of those configurations is realized in nature; and to it corresponds the particular set of masses of the vector bosons we observe.

The Higgs field really performs a miracle for electroweak theory. As a result of its introduction in the equation that governs the physics of subnuclear interactions, weak bosons automatically acquire the large mass required to make the theory consistent and its predictions calculable. And with the 2012 discovery of the Higgs boson at the CERN laboratories, the standard model is finally complete: one does feel that nature can be understood.

Beyond the Standard Model

Because of the elegance and economy of the symmetry breaking mechanism, the existence of the Higgs boson was hardly disputed. In a way, the predictability of the Higgs discovery was almost disappointing, as physicists love to be surprised. Anything new and unexpected can be used as a compass to guide the way into new unexplored territory. Further, there is a real need to find out what is hiding behind the beautiful construct of the standard model. Regardless of its beauty and the precision of its predictions, the standard model is incomplete and, to some extent, unsatisfactory: it cannot be the final word on the theory of matter and interactions. The model does not include a description of gravitational interactions, it does not provide a straightforward means to unify the four known forces together, and it presents at least one or two technical issues. Furthermore, the model is specified by two dozen parameters whose value is only known through experimental measurement, as there is no way to calculate them from first principles. The presence of over 20 unexplained numbers is annoying in a theory so beautiful and successful.

Among the parameters of the standard model, the ones we would most love to explain are the 12 fermion masses. It would be awesome if we knew why the muon is 200 times heavier than the electron! Note that the Higgs boson, which has interactions to fermions of strength proportional to the fermion mass, does provide an explanation of how mass manifests itself. It all works just as if quarks and leptons felt a drag of intensity corresponding to their different mass while propagating through empty space; the drag is due to their different interaction strength with the Higgs boson field. Yet, this only recasts the question in different terms: what is it that really causes different interactions with the Higgs field? The answer lies outside of the standard model and we have not found it yet.

We are thus led to believe that the standard model needs to be replaced by a deeper, more fundamental theory, and we can only hope that this new theory will give answers to all our questions, rather than answering just a few, while adding new, more impenetrable ones. In the remainder of this chapter I will briefly mention some of the more fashionable ideas of what that new theory might be.

Among the extensions of the standard model that have so far survived experimental scrutiny, *supersymmetry* (also known as SUSY) occupies a privileged spot. SUSY is a popular theory because it elegantly solves some of the shortcomings of the standard model. One of those shortcomings is the apparent *fine-tuning* of the standard model parameters, whose values conspire to produce a Higgs boson mass many orders of magnitude smaller than what it could "naturally" be expected. All existing massive particles modify the physical value of the Higgs mass by participating in quantum-mechanical "loop" processes: these are reactions whereby the Higgs boson turns, for a fantastically short instant, into a fermion–antifermion pair, or when it emits and immediately reabsorbs a W or Z boson. The resulting contributions to the Higgs mass are individually huge: some of them are negative, whereas others are positive; the puzzle is that their sum magically turns out to be tiny. A way to understand why fine-tuning is a big theoretical problem is to consider the following analogy, which was originally suggested to me by an anonymous visitor in the comments thread of a blog post I had written on the power and limits of the analogy in particle physics:

> "Suppose that sum of the profits and losses of ten companies turn out to total $10. If you only know this fact, and that *a priori* profits and losses in any given year appear to be more or less equally likely in the highly competitive market where these companies operate, what kinds of gross revenues do you expect the average company to have? Knowing only these facts, you might expect for a string of lemonade stands in a busy residential neighborhood. But you would be shocked if you learned that the companies had gross revenues of tens of billions of dollars a year each, and the total profit nets out to $10 simply by chance, even if profits and losses are equally likely. You would suspect that someone had carefully combed through tens of thousands of corporate reports to come up with a combination that was so equally balanced on purpose. Physicists, similarly, suspect that the loop corrections probably balance due to some hidden structure that we have not yet discovered, based upon the unnaturalness of the near perfect balancing in the absence of some unknown non-random principle that causes them to do so."

Explained with the profit–loss analogy, the puzzle is compelling: the cancellation is a strong hint of the existence of something fundamental that we have not yet understood. Now, the good thing about SUSY is that within its framework the corrections to the Higgs mass naturally cancel, thanks to (and at the expense of) the introduction of many yet-to-be-observed supersymmetric particles, or "sparticles," as well as a much larger set of free parameters.

If SUSY is the correct theory of nature, there must be a symmetry between fermions and bosons. Each standard model particle of half-integer spin must have a supersymmetric partner of integer spin; and similarly, to each standard model particle of integer spin, there must be a corresponding supersymmetric partner of half-integer spin. Quarks have *squark* partners, and leptons have *sleptons* counterparts. Each interaction carrier, called "gauge" boson, is also doubled up by a corresponding *gaugino*. As squarks, sleptons, and gauginos have not been discovered so far, the conclusion that SUSY is a beautiful but wrong idea can only be avoided by invoking a new kind of symmetry-breaking mechanism. If the symmetry between particles and sparticles is somehow broken, there is no need for SUSY particles to be as light as standard model ones; if they are very heavy (and thus hard to produce in particle collisions), that may explain why they have not been seen by experiments conducted so far.

SUSY production at a collider would create a quite rich phenomenology, whose details are, however, hard to predict, as they depend on the values of over a hundred unknown free parameters. Only a few general features are common to most of the possible realizations of SUSY; in particular, a weakly interacting, electrically neutral particle called *neutralino* is expected to be the final product of all decay chains of heavy superparticles. Neutralinos are the lightest SUSY particles: this makes them absolutely stable in theory realizations where the total number of sparticles is posited as constant, as they have then no way to decay into anything lighter. The production of neutralinos in particle collisions would yield no direct signal in the detector. They would leave without a direct trace of their passage, exactly as standard model neutrinos do. Yet, this would create a detectable imbalance in the sum of momenta of all visible particles produced in the collision. An example will clarify this point.

Suppose we observe two airplanes colliding head-on in flight. We expect the crash to throw debris in all directions; instead, we would be surprised to see all the pieces sent in a particular direction orthogonal to the common trajectory of the planes, with nothing visible recoiling against it. The surprise stems from the fact that an imbalance in the momenta of emitted debris perpendicular to the initial motion violates momentum conservation, a law we know by intuition even before we study it at school. In similar fashion, the colliding protons carry no net momentum in the plane perpendicular to the beams; that property must be retained by the particles that are created in the collision. An observed "missing transverse

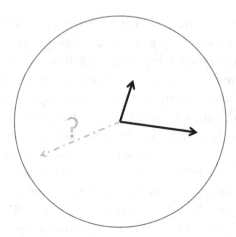

Figure 3.   Schematic "transverse view" of particles produced in a collision. The beams are orthogonal to the page and intersect at the center of the circle. Black arrows show the momentum vectors of two measured particles produced in the interaction. Since the two vectors do not balance off, there is missing momentum (see text). The production of an invisible additional particle emitted toward the left (dashed gray arrow) can then be hypothesized, to restore the balance.

momentum" (more frequently called *missed transverse energy,* a slight misnomer which has caught on among particle physicists) is the experimental footprint of the creation of neutralinos (see Figure 3). Note that the same behavior described above is typical of neutrinos, which can be produced, e.g., in the decay of W bosons; a significant amount of missing transverse energy must therefore be accompanied by other distinctive features of supersymmetric particles in order to provide a more striking signature of a SUSY process. Such additional evidence might come from the presence of a large number of collimated streams of hadrons (what we call "jets") produced by the decay of energetic squarks and gluinos. Jets are commonly produced by regular quarks and gluons, yet the production of many energetic jets accompanied by large missing transverse energy remains one of the most promising experimental signatures of supersymmetry at a particle collider.

The observation of a neutralino would be particularly exciting because the existence of that particle could solve the enigma of *dark matter* in the universe. Dark matter is a form of matter that does not clump into stars and emits no visible light. We know it exists since we observe its effect, e.g., on the measured rate of rotation of galaxies; yet, we do not know what it is made of. If the neutralino really existed, that particle would have been created in large numbers by the big bang, the explosion that generated our universe 14 billion years ago. Being as stable as protons and electrons, neutralinos would still be around today; unable to coalesce gravitationally and form luminous bodies, they would be classified as dark matter in the matter–energy budget of the universe. They would influence the dynamics of galaxies and galaxy clusters much as we indeed observe, if their mass were in the ballpark of a few hundred GeV. The detection of neutralinos would thus constitute a giant triumph for both modern particle physics and cosmology.

Other Fancy New Stuff

While SUSY is fascinating to many physicists, it is not the only idea on the market. A theory recently in fashion, and with a catchy acronym — LED, for *large extra dimensions* — proposes the existence of additional space–time dimensions beyond the four that we experience. In general, additional dimensions of space–time might provide an explanation of the weakness of gravity with respect to the other three forces. Just as a drop of strawberry syrup dilutes in a bucket of water and is hardly detectable to the eye once you stir, while a drop of oil only distributes on the surface and does not go unnoticed, gravity would appear weak to us, if we lived on a 3D "surface" of a higher-dimensional "bucket-universe," because it would dilute into the unseen dimensions. The other forces, confined on our "surface" like the drop of oil, would instead appear stronger. The new "directions" of space might be small or curved onto themselves, and thus undetectable by our senses. Yet, high-energy collisions might demonstrate their existence by producing processes whereby a particle "escapes" into an extra dimension, effectively disappearing from our myopic perspective. The signature of such a process would be similar to the one described above for neutralinos: large missing energy.

A nagging question that has haunted particle physicists ever since fermions got organized into three different generations is whether *three* is a magical number, or an accident of the limited energy range which we can explore with today's experiments. Is there a fourth generation of quarks and leptons out there, waiting for us to discover it, or even more such replicas? The large electron–positron (LEP) collider experiments in the 1990s appeared to answer the question conclusively by studying the decays of millions of Z bosons produced in 91-GeV electron–positron collisions. They measured a Z boson lifetime which only agreed with the existence of three kinds of neutrinos: more neutrino species would have enabled additional decay possibilities, making the lifetime shorter. However, soon thereafter, in 1998, the Super-Kamiokande experiment in Japan demonstrated that neutrinos have a tiny but non-zero mass. Consequently, the LEP indicium suddenly lost its grip. If all neutrinos were posited to be exactly massless, then the measured properties of Z bosons would imply that there are only three kinds of them; but since neutrinos do have a non-zero mass, one is free to hypothesize that there be additional neutrinos with a mass larger than half the Z mass. That would prevent them from being possible products of Z decays and it would leave the Z lifetime unaffected.

There is yet another possible way of extending our description of the subatomic world which is worth mentioning here. In the standard model, all fermions and bosons are described as point-like, dimensionless objects. But this may be an approximation: quarks and leptons might well be extended objects instead, endowed with an inner structure so far undetected due to the insufficient energy of our projectiles. A straightforward way to test the point-like nature of quarks is to study the most energetic particle collisions: if quarks are made up of still smaller entities, called "preons," the highest-energy collisions might manage to involve them directly; the scattering of preons off one another would give rise to an increase in the rate of such events.

The ones mentioned above are only a sample of the many ideas which have been put forward in the attempt of extending the standard model in the course of the last few decades. Unfortunately, none of them has yet received experimental confirmation.

Chapter 2

The Tevatron and the Collider Detector at Fermilab

In this chapter we focus on the physics of the hadron collisions of the Tevatron collider at Fermilab, and on the CDF experiment. In order to understand the design choices that were made for the CDF detector, a few key concepts for the complicated physics of hadron collisions must be briefly described. Among these are the *cross section* of subnuclear processes and the *luminosity* of particle accelerators.

Luminosity and Cross Section

Cross section is easy to understand, since it means the same thing for elementary particles and macroscopic targets. If you are at a firing range shooting at various targets set at the same distance, you will find it is easier to hit the largest ones — those with the largest cross section. Similarly, the cross section of a particle reaction is the effective area over which the intersection of the particle path results in the reaction. Being an area, the cross section can be measured in square centimeters; for subatomic particles, however, we use submultiples of that macroscopic unit. The basic unit is called *barn*, which owes its name to the sentence "as big as a barn." This sounds funny, as a barn is equivalent to a trillionth of a trillionth of a square centimeter; yet a barn is indeed big for particle reactions, as the cross sections of most subnuclear processes are still far smaller than that.

Fractions of the barn come handy: the *millibarn* (a thousandth of a barn), the *microbarn* (a millionth), the *nanobarn* (a billionth), and the *picobarn* (a trillionth).

Tightly connected to the concept of cross section is a crucial parameter of high-energy particle colliders: the *instantaneous luminosity* that those machines provide. Luminosity is a measure of how intensely the accelerator "illuminates" the interaction region with its particle beams. The higher the luminosity, the happier experimental physicists get, as denser crossings of particles produce a larger number of rare, interesting reactions. Since instantaneous luminosity describes how many particles cross a given area in a given time, its units are "counts per area per second." Hence, it can be measured in units of "counts per barn per second." The logic of this cryptic measurement unit will be clarified in a moment.

The instantaneous luminosity of a collider is not a fixed parameter: it may grow if improvements are made to the beam orbits, if the number of particles in the beams is increased, or if the collision region is shrunk. Otherwise, it typically decreases exponentially during a *store*, the colliding phase of accelerator operation. The decrease occurs because particles get continuously removed from the beams as they collide with their counterparts in the other beam or with residual gas molecules in the vacuum chamber where they circulate.

Running an accelerator at high instantaneous luminosity "strains" the detector collecting the data, because the collision rate is correspondingly high, more particles are produced, and the flux of electronic signals to be read out increases. The detector hardware is hard pressed to make the necessary selections and collect the most interesting events, which come in at larger rates. However, there is no other way to acquire as much *integrated luminosity* as possible. The integrated luminosity of a store, or of any given running period, is the sum over time of the values of instantaneous luminosities that occurred during the corresponding data collection period.

The importance of maximizing integrated luminosity can be understood by considering a couple of analogies. First of all, imagine you are a speleologist at the entrance of a huge unknown cavern. It is dark, and your lamp only allows you to illuminate the closest features around you. A more luminous lamp would allow you to see deeper into the chasm;

alternatively, you might take a long-exposure photograph. By "integrating" the scarce light bouncing off a distant wall of the cavern for a long time, you may capture faint details otherwise invisible to the naked eye. Similarly, a large integrated luminosity allows you to study more infrequent subatomic processes. You can get it by either running at high instantaneous luminosity (i.e., using a brighter lamp) or by collecting data for a long time.

Second, imagine you aim a machine gun at some target and start shooting at random toward it, showering the target with a flux of one bullet per second per square centimeter. Let us also say that the target has an area of 10 square centimeters. Given the above flux, in a minute you "illuminate" the target with an average of 60 bullets per square centimeter. By multiplying the target area of 10 square centimeters by that number, you easily compute that in a minute the hits will on average be 600. The concepts of instantaneous luminosity, integrated luminosity, cross section, and event counts all have a one-to-one correspondence with the quantities discussed above: the bullet flux corresponds to instantaneous luminosity; the target area is the cross section; the total bullets per area you shot are equivalent to integrated luminosity; and the total hits correspond to the number of produced events. If a collider integrates a luminosity L and the collisions produce a certain physical process that has a cross section σ, then the number of events of that kind will be σ times L. This is the most important formula in the physics of particle collisions!

While instantaneous luminosity is measured in counts per barn per second, integrated luminosity is measured in counts per barn, i.e., its physical dimensions are inverse barns, or multiples of that. An inverse millibarn, for instance, is 1000 inverse barns. The integrated luminosities to which the datasets discussed in this book correspond are much larger than that, so they need other suffixes: inverse picobarns and inverse femtobarns are the typical units of measurement.

Inside the Proton: The Physics of Hadron Collisions

The collision of two hadrons is a complex and very rich phenomenon because hadrons are themselves quite complex; they do not behave as point-like objects devoid of inner structure when we probe them with very energetic collisions. What we can treat as point-like are the constituents *within* the protons: these are the quarks and gluons, which we may collectively call *partons*. At high energy, a proton can be thought of as a stream of partons, and it is the latter, not the former, that produce an energetic reaction as they collide. Picturing protons that way brings up an important question: how much of the total kinetic energy of the system is released when a parton–parton collision takes place? This is important because new particles are created from the conversion of that kinetic energy.

Two theoretical physicists, James Bjorken and Richard Feynman, studied this problem somewhat independently and came up with a successful model in the late 1960s. Their studies focused on a variable now known as "Bjorken *x*," the fraction of the proton momentum carried by the parton. If we knew the momentum fractions carried by the two colliding partons, we could compute the energy of the parton–parton collision.

In order to determine the probability that our collision involves partons of given momentum fractions we need to embark in a preliminary study, which will bring us to know better our projectiles and their internal structure and composition. Imagine the proton is a roughly spherical black bag full of unknown stuff (the partons that make it up), and you cannot open it to look inside. You could get a rough idea of what is inside the bag by throwing something at it. You would not want to hit it with another proton (another parton bag) since that would just add more mess to an already messy situation. You would probably use something very small, very hard, and very fast. You could then follow the projectile's trajectory as it enters the bag, kicks something out, and emerges scattered at some angle from its original direction. What I am describing is called *deep inelastic scattering* (DIS), an experimental technique which is usually performed with electrons or muons aimed at a nuclear target. DIS has been investigated in great detail by a number of experiments in the last 50 years.

As we saw in Chapter 1, the quark model posits that a proton contains three *valence* quarks: two up quarks and one down quark. It is a useful schematization, as valence quarks determine the intrinsic, static properties of the proton. However, to really understand the dynamics of the system, we need to consider a significantly more complex description. DIS experiments have shown that the proton contains also gluons and virtual quark–antiquark pairs of all flavors, which constitute what is called the *sea*. Virtual quark–antiquark pairs continuously fluctuate out of and back into the vacuum, borrowing from it the energy necessary to come into existence for time intervals so small that the temporary imbalance is inconsequential.

Protons are made by valence quarks, sea quark–antiquark pairs, and gluons. The strong interaction entirely determines their dynamics: understanding QCD is therefore crucial if we want to predict phenomena involving protons (and all hadrons in general), such as the relative rates of different reactions. Unfortunately, we are not entirely there yet! We can compute the rate of high-energy processes, but what happens at low energy cannot be precisely determined with the tools at our disposal.

Partons move almost freely inside the proton, undergoing low-energy interactions. These are the ones we cannot calculate. Hence we cannot predict how the interior of a proton works; but we can still measure reactions that give us a pretty good picture of it. We rely on DIS measurements to experimentally determine the behavior of partons. This allows us to estimate the probabilities of finding given quarks or gluons inside a proton, as a function of the fraction of proton energy that the constituents carry. These probabilities, called *parton distribution functions* (PDF) determine the relative frequency of processes that release different amounts of energy. By "factorizing out" the part that we cannot calculate, we may then focus on the physics of the collisions.

Armed with the above information, we may try to describe a hadron collision as it develops. Two constituent quarks or gluons, one from each proton, find themselves on the same trajectory and hit each other head-on. This is the pair of partons that ends up producing most of the action. Whatever is left of the two projectiles will uneventfully break up and recombine without giving rise to exciting physics. The hard-hitting duo is called the *initial state* of the collision, and they originate what is called the *hard subprocess*, or the high-energy reaction. Each of the two constituents usually carries only a small fraction of the energy of their parent; those fractions determine the total energy of the subprocess. The result of the reaction is a *final state* made up of new energetic quarks or gluons that are knocked out of the interaction region. Eventually, a process called *hadronization* recombines the final state partons into colorless hadrons. The latter constitute the observable outcome of the collision: the physical reactions that preceded their formation can be guessed by studying their dynamics.

Just as PDF cannot be calculated, neither can the mechanisms responsible for the creation of new hadrons from the final state of the hard subprocess. This should be no surprise, as both processes are governed by a large number of low-energy parton interactions. The process starts with a phase called *parton shower*, when the energy of the quark or gluon kicked out of the interaction point starts to decrease, through the emission of low-energy gluons and the splitting of gluons into quark–antiquark or gluon pairs. There follows a *hadronization* phase, when partons recombine in colorless hadrons. Eventually, from one single energetic parton emitted by the hard subprocess, a stream of hadrons roughly traveling in the same direction is created. It is such a stream of hadrons what we usually call a *jet* (see Figure 1): the observable result of a parton kicked out from a proton is the emission of a score of light hadrons sprayed around the original quark direction, within an angular spread of few tens of degrees.

The non-calculability of the hadronization phase is a hindrance to the extraction of detailed predictions of the process of jet creation, which are

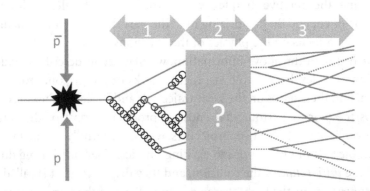

Figure 1. Three phases of formation of a hadronic jet from an energetic parton created in a hard collision. When a proton–antiproton collision takes place (left), a quark (full line directed to the right) can be emitted with high energy. In phase 1, which we can compute with QCD, a parton shower is generated: the quark radiates a gluon (pictured as a spring) and the latter emits another gluon; gluons may split into further quark–antiquark (full lines) or gluon pairs, and quarks can radiate other gluons. Then the hadronization of quarks and gluons (phase 2) takes place through non-calculable low-energy phenomena. Finally (phase 3), the produced charged and neutral hadrons (respectively, full and dashed lines, on the right) decay into stable particles, forming a hadronic jet.

needed to really understand hadronic collisions. But physicists do not despair when they encounter non-calculable phenomena. They know how to get around it: they *model* the physics. A model is no more than a simplified mathematical description of the mechanisms at work. It is usually not meaningful or profound, but it has a practical value as it allows us to obtain a good approximation of the result of a physical process.

Quark and gluon hadronization models contain a few tunable parameters which one can control to match one's predictions to the observed data. A good hadronization model may reproduce things such as the fraction of energy carried by the highest-energy hadron in the jet, the total number of produced particles, and other details. Since hadronization is a stochastic process, where the same initial conditions may give rise to different outcomes, the model must be encoded in a *Monte Carlo simulation*. This is a computer program that generates random numbers to calculate the model's predictions. The Monte Carlo method works by randomly choosing the value of each of the variables describing the process under study. The choice is performed by following the model of the relative probabilities that those values occur. If the model is good, repeated random sampling will approach a good representation of the process.

The Monte Carlo method is universally used in particle physics. Simulating the expected characteristics of particle collisions recorded in a detector requires a computer program to describe the effect of many different physical processes: the interaction of the colliding bodies, the hadronization and decay of the produced particles, and the interaction of the stable particles with the detector components. Any differences between real and simulated data can be studied to verify whether the cause is a flaw in the model used by the simulation, or new physics of which the simulation does not know about.

The Tevatron

The history of the *Tevatron collider* may be traced back to 1972, when particle physicists at Fermilab built a synchrotron called the "main ring," a proton accelerator hosted in a 6.3-km long circular tunnel dug a few meters underground. The main ring was capable of imparting protons with an energy of up to 400 GeV, an order of magnitude higher than until then achieved. The Tevatron produced the data where in 1977 a team of physicists led by Leon Lederman discovered the *Upsilon meson* (usually labeled with the Greek letter Y), a bound state of bottom–antibottom quarks. Three years after the confirmation of the quark model, the existence of the Y showed that quarks come in three generations. In the meantime, the CERN Laboratory was commissioning the *SpS*, a proton synchrotron capable of reaching 315 GeV of beam energy, and there were already thoughts of converting that machine from the usual *fixed-target* setup into a proton–antiproton collider. This would drastically increase the energy available for new particle production. In a fixed-target setup like that of the main ring or the SpS, the particle beam is directed to a stationary element. Because of momentum conservation, the final state of each collision must possess large kinetic energy in the original direction of the projectile. This reduces to a small fraction the energy available for creation of new particles. Indeed, the 315-GeV fixed-target collisions of the SpS could release at most 24 GeV of energy; this would become $315 + 315 = 630$ GeV in a collider setup!

The CERN competition at the high-energy frontier demanded a response from the USA. For a while, the idea of Fermilab's director Robert Wilson was to construct a "doubler," a synchrotron using for the first time the technology of superconducting magnets which Fermilab was developing. The higher bending power of the strong fields generated by those magnets could allow a proton beam circulating in the same tunnel of the main ring to reach the unheard-of energy of 1 TeV (i.e., 1000 GeV). Once such a machine were built, one could imagine bringing the protons of the doubler to collide with the protons of the main ring. However, such a project entailed non-trivial technical difficulties. The ongoing development of new techniques demonstrating how to obtain narrower and more focused particle beams finally suggested that the right machine to build was a proton–antiproton collider, similar but more powerful than the one that CERN was considering.

Using antiprotons in a collider was a daring challenge. High beam intensities can be obtained with relative ease in a synchrotron using protons, which are readily available as ionized hydrogen atoms. For proton beams the limiting factor determining the total intensity is the capability of the acceleration system to counter the repulsive electrostatic force of densely packed particles with *ad hoc* focusing systems. Several additional electrodynamic effects, of amazing complexity in such a big machine, must be kept under control, but conceptually they are no impediment. On the other hand, antiprotons need to be created from scratch, painstakingly accumulated, and stored safely to prevent their premature annihilation with ordinary matter. For this purpose, one needs to direct an energetic, intense proton beam toward a fixed target, and select with a suitable setup of bending magnets the few antiprotons generated by the ensuing collisions. As an antiproton is created less than once in 10,000 collisions, the creation of a dense and focused antiproton beam is a formidable problem. Yet, that was the way to go if the laboratory was to continue along the path of leadership in particle physics research it had convincingly taken in recent times. Provided that an antiproton factory were designed and built, the same synchrotron and magnets could accelerate and bend antiprotons in the opposite direction of protons, creating collisions at the fantastic energy of 2 TeV!

Two-TeV collisions were a broad stride into unknown territory, calling for the construction of a multipurpose detector capable of detecting whatever new physics signal could be hypothesized. Alvin Tollestrup, who had just set back in track the project of superconducting magnets that would instrument the Tevatron ring, was appointed as the head of the detector project. It was quite clear to him from the start that an international collaboration was needed: Fermilab would greatly benefit from the experience in detector building that a few strong groups of physicists coming from different countries could offer. Visiting Tokyo for an international conference, Tollestrup convinced Kunitaka Kondo, a professor from the University of Tsukuba, to join forces toward the construction of a detector for the Tevatron collider. This was the start of a quite significant and long-standing contribution from Japan in terms of funding, manpower, and ingenuity. The Italian Giorgio Bellettini was similarly invited by Tollestrup to join the Americans and Japanese. He took part in the project at the start of 1980 with his colleague Paolo Giromini. The duo soon attracted several other physicists from Pisa and Frascati; other Italian universities joined the project in the following years.

The detector project was dubbed CDF, for Collider Detector at Fermilab (originally "Collider Detector Facility"). The challenge presented by the construction of a new detector was multifold. CDF was to explore an entirely new energy range, to measure W and Z bosons, and to search for possible signals of new physics. Those tasks posed specific construction demands: a strong magnetic field to measure charged particle momenta of 100 GeV and above, and a thick *calorimeter*, a detector capable of entirely absorbing and measuring hadronic jets of several hundred GeV. The new facility would not become operational in time to compete with CERN in the rush to find W and Z particles (which were in fact discovered by the UA1 experiment at the *Sp$\bar{p}$S*, CERN's proton–antiproton collider, in 1983), but it would be well positioned to search for new physics phenomena at energies that the CERN machine would not be able to reach.

The Design of the CDF Detector

There were initially several competing plans on how to design the detector. In the end, the one which won over the project leaders relied on well-understood detection techniques, whose performance and reliability could be trusted. An important, desirable structural feature was that the detector be as hermetic as possible. A proton–antiproton collision of high energy typically produces hundreds of elementary particles flying out of the interaction point in all directions. Detecting only a subset of those particles makes it harder to understand the nature of the subatomic process that originated them. However, a complete hermeticity of the detection apparatus is impossible to obtain, since the proton and antiproton beams need to enter and exit from a hole somewhere! Also, there are high-voltage cables that deliver power to the detector and electrical wiring that carries the electric signals out of it. Gas constantly needs to flow into and out of *drift chambers* (the wire detectors commonly used for charged particle tracking), and heat must be removed from the electronic components. A complete detection system is quite complex and making it hermetic is a formidable challenge.

One may consider two possible near-hermetic layouts for the detectors around the *beam pipe*, the vacuum tube where the beams circulate: a cylinder around the beam closed by endcaps on the two sides, or a sphere pierced through one diameter by the beam. From an engineering standpoint a cylinder is easier to build than a sphere, because its axial symmetry allows for concentric layers of sensitive elements which are straight along one dimension. The support structures of a cylinder are also easier to design. A cylinder offers another important advantage if one considers the physics of proton–antiproton collisions. The generated particles that matter the most are those carrying a large momentum in the direction orthogonal to the beam axis. They are the most important ones to measure precisely if one wants to puzzle out the details of the hard subprocess, as they are the real messengers of the hard interaction. Their *transverse momentum* (labeled p_T) directly measures the intensity of the force which accelerated them off the collision point. The same does not apply to particles with a large momentum component in the longitudinal direction, which is affected by the unknown motion of the center of mass of the colliding partons. Hence a cylinder, which allows for a privileged treatment of the direction orthogonal to the beam axis, is to be preferred. Finally, a cylinder may naturally integrate a solenoid, which provides a uniform magnetic field in its interior, capable of bending the trajectories of charged particles emitted by the collisions and thus making their momentum measurable.

In the minds of its creators, the new detector had to not only identify and measure with precision the energetic charged leptons emitted by W and Z boson decays, but also reconstruct the jets produced by the hadronization of quarks, inferring the energy of the originating partons. The latter was a crucial ingredient if CDF was to hunt for new phenomena at the high-energy frontier. Such breadth of objectives drove the design toward a multipurpose system providing a large amount of redundancy. Redundancy is crucial when dealing with ethereal bodies such as elementary particles, which we cannot see, touch, or come back to measure again if we are not sure of our first result. By obtaining independent information with different methods, one may perform crucial cross-checks of the robustness of the measurement. Systematic uncertainties can then be reduced, improving the performance of the apparatus. Redundancy also has another merit: one is not putting all the eggs in the same basket. If a detection system underperforms, others can usually make up for most of the deficit, keeping the overall performance of the apparatus almost unaffected.

In the end, the design of the CDF detector was a textbook example of a multipurpose, redundant, and precise system tailored to the features of hadronic collisions of the highest energy ever achieved. It featured an inner tracking system capable of detecting charged particles and determining their momenta with great accuracy from their curvature in the magnetic field generated by a superconducting solenoid. Outside the magnet coils was placed a segmented calorimeter system, with an inner section suitable for the measurement of electromagnetic showers and an outer one where hadrons could yield their energy to showers of nuclear interactions. The central calorimeter was in turn surrounded by eight layers of muon detectors to measure the trajectory of muons, the only detectable particles capable of punching through large amounts of heavy material.

The Components in Detail

The central tracking chamber (CTC) was a cylinder with a radius of 1.2 m and a length of 3 m. Filled with an argon–ethane gas mixture, the CTC provided precise position measurements of the charged particles originating at its center. The physical basis of the reconstruction of track positions in such a device is simple: a charged particle traversing the gas ionizes a few dozen atoms per centimeter along its path, by knocking off electrons. The electrons are prevented from recombining with the ions by a strong electric field, which forces electrons and ions to drift in opposite directions. The field is generated by wires of opposite polarity and is shaped in such a way that the drift velocity remains constant until electrons get close to the anode wire. In the proximity of the wire the field rapidly increases, so electrons gain kinetic energy and produce secondary ionization charge. The resulting electrical impulses, suitably amplified and encoded, give information on the position of the ionizing track.

The CTC was embedded in a superconducting solenoid which produced a uniform magnetic field of 1.4 Tesla. This was enough to appreciably bend the trajectories of charged particles even at the highest momenta, allowing a measurement of momentum from its proportionality with track curvature. In fact, charged particles follow helical trajectories inside a constant magnetic field, due to the action of the Lorentz force. A particle with a momentum of 100 GeV moving orthogonally to a 1.4-Tesla magnetic field follows a circle with a radius of about 250 meters. With a simple calculation, one finds that from the center to the boundary of the tracker the particle trajectory bends by about two millimeters. Since the uncertainty on track position amounts to a tenth of that amount, particle momenta of 100 GeV can be measured with 10% precision.

The *electromagnetic calorimeter* system was a cylinder built around the solenoid in a typical sandwich-like design, radially alternating layers of 5-mm lead sheets and 1-cm sheets of *plastic scintillator*. Electrons traveling at close to the speed of light through the strong electric field of lead atoms radiate energetic photons by the process called *bremsstrahlung* (braking radiation). These photons may then materialize new electron–positron pairs as they interact with other lead atoms. The multiplication process continues in a shower-like progression until the energy of the generated photons becomes too small to create further electron–positron pairs. Traversed by the shower, the plastic scintillator embedded within the lead sheets releases ultraviolet light, which is finally converted into a usable electronic signal by photomultiplier tubes. The number of particles in the shower is proportional to the energy of the incident electron, thus the signal provides a measurement of the total released energy.

Installed around the electromagnetic calorimeter, the *hadronic calorimeter* used a similar, but bulkier sandwich design of iron layers alternating with scintillating material. Hadrons traveling inside dense matter lose most of their energy in strong interactions with heavy nuclei. But strong interactions are confined within the radius of hadrons, so it is really a hit-or-miss situation. Since the atomic nucleus is very small, a hit occurs infrequently: the so-called *nuclear interaction length* (the typical distance traveled before an interaction takes place) of iron is 17 centimeters. When a collision happens, it usually destroys the original hadron and creates many secondary particles, each carrying a small fraction of the energy of the parent body. These secondary hadrons find other nuclei to break apart after traveling another nuclear interaction length (on average). The multiplication process goes on until there are energy to lose and nuclei to hit, resulting in what is called a *hadronic shower*. Hadronic showers are longer than those produced by electrons and photons, but the method by which they are measured is the same: energy is proportional to the released scintillation light.

In summary, the momentum of all charged particles is precisely measured in the tracking chamber; the energy of electrons and photons is determined by the electromagnetic calorimeter; hadrons are measured in the hadronic calorimeter. But then there are muons. Muons are extremely interesting products of hadronic collisions, as they signal that a weak interaction has taken place; that may be the decay of a weak W or Z boson, or a heavy quark. And muons are special since they lose very little energy in electromagnetic interactions with atoms. This allows them to traverse unhindered the heavy material of the calorimeters. It is therefore profitable to surround calorimeters with muon detection elements: a signal there, matched to a track reconstructed in the central tracker, makes a clean muon candidate.

Four layers of drift chambers, called the central muon system (CMU), were placed around the central calorimeter structure. Each chamber was a long cavity of rectangular section. At the center lay a thin wire brought to high electric potential. The cavity was filled with gas which could be readily ionized by the passage of a charged particle; the ionization signal was detected as explained earlier.

The CMU reconstructed muon tracks well, but it occasionally received residual particles from showers in the hadronic calorimeter. Light hadrons of high momentum could punch through the detector and be mistaken as muons. From 1992 on, a 60-cm-thick steel shielding was placed around the CMU to filter out this background noise, and additional muon chambers were mounted on top of it. This outer system was called central muon upgrade (CMP). The coincidence of CMU and CMP muon chamber signals effectively filtered out the hadronic background, making the identified muons trustable tags of interesting collisions producing W or Z bosons, b-quark jets, or top quarks.

Putting it Together

The CDF detector components described above correspond roughly to what existed in 1988, when the Tevatron first delivered data in significant amounts. The subsystems were constructed in nearby sites and then put together in the assembly hall, a pit dug inside the B0 building. The floor of the assembly hall stood some 10 meters underground, at the same level of the adjacent collision hall which was traversed by the Tevatron beams. The central part of the detector could be moved from assembly to collision hall on two large rails; the two experimental areas were physically kept separate by a movable concrete wall which screened the high radiation produced during the proton–antiproton collisions.

Building the central structure of the detector was a demanding engineering challenge, which took several months in the course of 1984 and 1985. The bulkiest part of the central detector, which determined its overall appearance, was the calorimeter system. This was made up by two large 24-wedge wheels; each wedge covered 15° in azimuthal angle. The wheels were placed concentrically with the beam line around the nominal collision point. The wedges were constructed at the Argonne National Laboratories, carried to Fermilab, and set around the solenoid.

Another instrument, called vertex tracking chamber, was installed inside the solenoid. It was filled with gas, had wires arranged in planes at right angles to the beam pipe, and it measured the interaction vertex position along the beam. The CTC, which would be crucial to precisely track the trajectories of charged particles and determine their momenta, was only installed after the first engineering run of the Tevatron, which took place in the second half of 1985 (see Chapter 3).

Once assembled, the central detector looked like a big and bulky black cube. After 1986, two "endcaps" would find their place in front and behind it. The endcap sections included electromagnetic and hadronic calorimeters capable of detecting jets produced by quarks or gluons emitted at small angle from the beam direction. The endcaps were made by two halves that sandwiched the beam pipe running at their center. They could easily be moved sideways for servicing tasks. A picture of the central part of the CDF detector is shown in Figure 2.

One in 50,000 Makes It

If the detector elements described above constituted the vital organs of CDF, the counting room was its real brain: most of the electronics handling the detector output resided there. Custom-made electronic boards provided for the digitization of the signals and the *triggering* of the data acquisition. The trigger was the system in charge of determining whether each detected collision was worth saving to magnetic support or to be discarded. That was a crucial task! The 5000 metric tons of lead, iron, tungsten, and steel would have been worth next to nothing if they had not been complemented by a system capable of taking that decision on the fly. Let us see why.

Figure 2. The central part of the CDF detector in the assembly hall. Image credit: Fermilab.

The rate at which collisions took place in the core of CDF was equal to one every 3.5 μs. The Tevatron beam was "bunched" into six packets of protons and six packets of antiprotons; these intersected in the core of the detector 286,000 times a second. The transverse size of bunches was designed to be small enough and the number of particles in each bunch large enough that every crossing had a significant chance of providing a proton–antiproton interaction. Yet, there would be no way to save at such a tremendous rate the information read out by the more than 100,000 electronic channels comprising the CDF detector: the data storage system was capable of recording the information of no more than about 50 collisions per second! But that was not a problem, as most proton–antiproton collisions were indeed not interesting. This can be clarified if you recall the bag analogy of the proton which we used earlier: most of the times, two bags hitting each other at high speed will result in a simple bouncing of the bags off one another. More rarely, they will break apart but still not show any hint of an energetic collision between their hardest constituents. Only in exceedingly rare instances will their dissociation be accompanied by some collimated debris violently sent in opposite directions: this is the signal of the production of an energetic collision between, say, two glass bottles contained in the bags. While the more common interactions between the bags tell us very little about their internal structure, the glass pieces occasionally produced in the hardest interactions tell us many details of the internal distribution and content of the bags.

Going back to protons and quarks, the vast majority of collisions are not very valuable, because they only produce a low-energy interaction. In order to get what we paid for with such a high-energy collider, we need to study the highest-energy collisions. These produce massive bodies such as W and Z bosons, or top-quark pairs, or even more exotic states of matter, if they exist. It really is enough to collect 50 events among the 286,000 produced per second, if we are capable of picking the ones that might tell us about new physics, or at least allow us to carry on measurements of heavy particles which were not possible in previous experiments. The rest of the collisions involve physical processes that lower-energy machines already investigated in detail in the past.

To make the above point clear, imagine you are digging holes in the ground in order to find out about the geological history of a piece of land. We assume that there has been a horizontal stratification of the ground in layers deposited during successive historical epochs, so the information we obtain by collecting samples at different depths is roughly the same wherever we dig. If we have studied with many 10-meter-deep holes the ground at various locations, finding everywhere the same pattern of layers of underground rock, then the first time we manage to dig a hole deeper than 10 meters we go straight to the samples of soil extracted from the deepest layers. We already know everything about the more shallow layers: a quick look suffices to verify they are similar to those of the other holes. Similarly, the trigger at a high-luminosity hadron collider is designed to select the rarest, most energetic events, "digging as deep as possible." Less energetic events do not teach us anything we do not already know.

The trigger was constructed in gradual steps when the detector was already in place, and it became fully operational only when Run 1 started, in 1992. It was organized in three logical levels, which physically corresponded to three floors of the B0 building.

The trigger rooms were located above the cavern housing the detector during data collection. The Level-1 trigger, located on the first floor, received the signal cables directly from the detector underground. In less than 3 μs, the system did the following: quickly determined (with coarse granularity) the deposited energy from the fast calorimeter signals, reconstructed segments of muon tracks from the muon detectors, and estimated their transverse momentum by the tilt of the segments with respect to the radial direction (less tilted segments indicated particles that were bent less by the inner magnetic field and thus had greater momentum). Also, a rough identification of energy deposits in the electromagnetic calorimeter allowed the spotting of potential electron signals. Using this information, the Level-1 trigger could decide whether the event deserved a closer look. If a sufficiently large amount of energy was observed, or if the signal of a muon or an electron above a certain momentum threshold was present, all the data from the detector was read in and passed to the second level. Otherwise, the event was discarded.

The Level-2 electronics had more time to take a decision about the several thousand events per second accepted by Level 1. Custom-designed electronics reconstructed the charged tracks, jets, electrons, muons, and photon candidates. A correspondingly more complicated list of possible selection criteria was used by the Level-2 trigger, which employed all measured characteristics of the objects contained in the event. The selection was tuned to accept and pass to the third level no more than 300 events per second on average.

Finally, the Level-3 trigger was an array of commercial processors running the full event reconstruction software. The programs performed a complete measurement of all the detected particles. *Hits* in the tracking chamber (points where the particle left a signal) were fit to helical trajectories to determine particle momenta. Energy deposits left by streams of hadrons in the calorimeter were measured and collectively reconstructed as signals of hadronic jets. Electrons and photons were identified using the corresponding signals in the electromagnetic calorimeter. Muons were identified as charged tracks pointing to hits in the outer muon chambers. A confirmation of the decision taken by Level 1 and Level 2 was possible with the now better-measured event characteristics. A total of less than 50 events a second were flagged to be written to media storage by the data acquisition system. The data were sent to different output streams, depending on their characteristics.

It is worthwhile to stop and think about the system, which allowed the drawing of three-dimensional pictures of the produced subatomic particles by reading out and interpreting in real time millions of electronic signals, several hundred thousand times per second. The formidable complexity of the CDF detector and its readout were indeed remarkable!

Chapter 3

Revenge of the Slimeballs

In late August 1985 a fixed-target run of the newly commissioned Tevatron was terminated ahead of schedule due to the failure of a magnet. A few days later, the 5000-ton central detector was rolled into the collision hall: an engineering run in collider mode could finally start. For the first time, the Tevatron began circulating protons and antiprotons at 800 GeV per beam, attempting to create collisions in the core of CDF. This was little more than a first test, meant to verify the operational status of the accelerator. It was not expected to deliver a significant amount of data to the still incomplete detector. Hence, there was no real chance to produce physics results yet.

A crew of excited and hopeful "regulars" led by the spokesperson Roy Schwitters populated in those days the CDF control room, at the second floor of the B0 building. The researchers nominally took eight-hour shifts, alternating in the operational tasks required to run the experiment; yet, they usually lingered around for much longer. Everybody hoped that the Tevatron would finally manage to produce the first proton–antiproton collisions.

23 Events

Toward the beginning of October, the CDF scientists started to show signs of distress. The Tevatron had been circulating low-luminosity proton and antiproton beams in the machine for a while now, but getting the

41

micrometric-sized beams to intersect in the center of the detector appeared to involve more black magic than straightforward science, since beam orbits in the brand new accelerator were still being tested. A synchrotron is a complicated system: as protons travel in its interior, they constitute an electric current. This current induces currents on all metallic elements surrounding the beam. The mutual interaction of those currents causes electromagnetic resonances that may disturb the orbit of the particles. The basic physics determining those effects can be found in many textbooks, but the dimension and the complexity of the machine make the search for stable and well-controlled orbits a trial-and-error game.

All what had been recorded by CDF until then were "beam–gas interactions," where a proton or an antiproton hits the nucleus of a residual gas molecule in the beam pipe; the vacuum of the latter was very good, but not perfect. Those collisions were quite characteristic: like a truck running at 100 miles per hour into a parked car, the proton sprayed debris exclusively in the direction of its original motion. Even without complex algorithms, and using visualization tools that nowadays appear prehistoric, it was possible to recognize that distinctive topology.

By the evening of October 12 time was running out. At 8 AM of the following day the 1985 run would terminate, as the machine was scheduled to go into a shutdown to allow for planned improvement work on the magnets and on the whole system. At midnight, the shift ended for Giorgio Chiarelli, a young researcher from Pisa. He was about to go home with a sense of failure; hour after hour in the last few weeks he had wished he would finally see the first proton–antiproton collisions appear on the screens of the control room, and now it was time to forget about it until the start of the 1986 run. But his co-shifter and colleague Sergio Bertolucci suggested that he not leave yet.

> "Why don't you stay? The machine guys are going to try one last time in a few hours. We might get lucky!"

Giorgio stayed, inspired by his colleague's optimism. And he did not regret it; as the next store started and particle bunches were injected in the Tevatron, the unmistakable sign of proton–antiproton collisions started to appear on the displays. Unlike the heavily broadcasted start-up of the Large Hadron Collider (LHC) in 2008, that night in the CDF control room there were no journalists. There were only a bunch of tired, but deliriously happy and excited physicists, who observed their detector finally coming to life and collecting collisions at record-high energy, as they had dreamt for months.

Bob Kephart and James Bjorken extracted from the graphs a down-and-dirty measurement of the distribution of charged particles that were being recorded by the vertex time projection chamber (VTPC), the only tracking device installed until then in CDF. "See, it's 4.5 tracks per unit rapidity! It's exactly as I computed it!" was Bjorken's enthusiastic claim once they were done measuring the tracks in the first recorded event. Bjorken was one of the fathers of QCD: if he was convinced that those were true proton–antiproton collisions, there was room to rejoice! In the matter of half an hour after recording the first collisions, the control room was brimming with enthusiastic faces: very soon the experiment leader Alvin Tollestrup arrived; then many others, including the lab director Leon Lederman. Event display printouts were circulated and hung on the walls.

A Polaroid picture of "channel 13," the TV display which logged the machine status, was taken and pasted on the logbook where all the relevant information about the detector operation used to be recorded. The display had a telling entry at 0310 (3.10 AM): "CDF reports a confirmed proton antiproton event. Look for free bubbly at B0." While sparkling wine was distributed in plastic glasses, the screenshot was signed by the physicists. That logbook entry became a very important page in the history of the experiment (see Figure 1). In the course of that last night of the 1985 run the machine collided protons and antiprotons at the total energy of 1.6 TeV, pulverizing the previous record of 630 GeV achieved three years earlier by the Sp$\bar{\text{p}}$S collider at CERN. The small number of particles in the beam yielded an extremely low luminosity: only 23 collisions were recorded! Yet, those collisions constituted an important milestone, a proof of principle, and a promise of better things to come.

The Sacred Sword

The 23 collisions collected by the detector in 1985 were an injection of optimism in the veins of the collaborators. CDF would soon be ready to

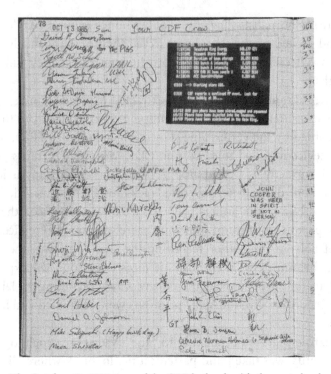

Figure 1.   The October 13, 1985 page of the CDF logbook with the pasted polaroid image of channel 13 (see text) and signatures of all the physicists present in the control room. Image credit: Fermilab.

compete with the other particle physics experiments around the world and, in particular, with the CERN experiments at the Sp$\bar{\text{p}}$S collider. That was a quite strong motivation to get the detector ready for a real data collection run. Everybody wanted lots of data. But CDF still needed to be finalized in a number of ways: most notably, by installing the endcap calorimeters, which would allow the measurement of jets emitted at low angle with respect to the beam direction; and by inserting in its core the central tracking chamber — the marvelous three-meter-wide detector on which CDF relied to precisely measure charged particles. Further, the electronics for data acquisition were not in place yet, and a proper triggering system existed only in blueprint.

In addition to the completion of the detector and its electronics, the detector elements already in place and working were in need of a precise calibration. Among other things, the response of the hadronic calorimeter had to be monitored and equalized. This was done by exposing each of its modules to a reference signal. Systematic differences in the total light yield of the stacks of scintillating sheets invested by ionizing particles of fixed, well-known energy could be recorded and later used to correct their output. The procedure consisted in inserting a long hollow rod of steel inside a small dedicated hole which ran through the length of each detector module. The rod was filled with a radioactive powder, a cesium isotope (^{137}Cs) which emitted energetic gamma rays. The gamma rays converted into electron–positron pairs in the iron slabs of the calorimeter, yielding a reference signal; the collection of that signal and the comparison of its intensity across the various detector elements allowed the extraction of the required inter-calibration values. The procedure was effective, but its practical execution was technically challenging, due to the difficult operation of the flexible steel rod.

The rod was indeed a rather odd thing, and it duly deserved the resounding name it had been given by the Italian physicist Aldo Menzione: "la spada sacra," the sacred sword. Like any other radioactive source, it required careful handling, a detailed bookkeeping of the duration of its use, and the deployment of clear signs to mark out the area where it was used. Radiological operators had to carefully plan in advance their use of the source in order to receive as small a dose of radiation as was reasonably achievable. Those safety rules were, however, arguably insufficient for a long and flimsy steel rod which emitted gamma rays along its whole length, made slippery and flammable by its graphite coating.

Two Italian collaborators, who worked at the calibration of the calorimeter in the summer of 1986, were the protagonists of one of the worst radiological incidents at the lab in those years. Fortunately, the episode can be now remembered for its rather funny twists rather than for its consequences. Standing on top of a cherry-picker (a movable elevated platform) in the CDF assembly hall, the two scientists together held the sacred sword, which was over three meter long and thus hard to maneuver. They directed it into the dedicated hole of a calorimeter module, pushing it through the full length of the detector; then they measured the current produced in the photomultiplier tubes by the scintillation light, extracted the rod, and moved to a different module. Admittedly, the setup was unusual. Already the simple operation of the elevated platform several meters above the ground was a peculiar sight; add to that the handling of a flimsy rod filled with dangerously radioactive material, and you get something closer to a movie scene by Laurel and Hardy than a particle physics test.

At some point the two Italians, who were concentrating their attention on the tip of the rod which they tried to insert in the appropriate hole, inadvertently caused an elastic movement of the other end of the rod. The rear tip of the sacred sword touched the terminals of the electrical battery powering the vehicle, lodged at its base. The ensuing short-circuit ignited the solder cap of the hollow rod. The cesium powder, released from the interior, got vaporized or dispersed on the ground of the assembly hall. Horrified, the researchers immediately extinguished the fire and rushed to the restrooms to wash themselves, thereby further spreading the contamination, and only then called the Fermilab security.

Witnesses of the incident swear that the scene outside CDF soon became one similar to the aftermath of a terrorist attack. In a few minutes, a dozen vehicles including fire brigade teams, police patrols, and ambulances crowded the area around the big orange industrial building, filling the air with loud sirens and bright-colored flashes. For the following few days, the assembly hall remained off limits, to decontaminate the area where the incident had taken place. In particular, the walls of the bathroom where the two Italians had washed themselves had to be broken down in order to dig out the sink tubes and recover the contaminated fluids.

As for the two terrified researchers, they were taken to the Argonne Laboratories, 20 miles southeast. At Argonne, in a underground facility designed to be screened from cosmic-ray backgrounds and radioactive contaminants, they could be analyzed to determine the radiation dose they had been exposed to. But they were not the only ones who suffered that treatment: a young colleague from Frascati who had not participated in any way to the incident was taken hostage and subjected to the same procedure, as the radiologists needed a "control sample." A few months before the incident, prior to their trip to the USA, the Italians had all been exposed to the radioactive fallout from the catastrophic Chernobyl nuclear disaster; because of that, a dose of radioactive contamination was expected in their bodies. In order to determine how much cesium had been absorbed from the vaporized source by the two researchers, a subtraction of the radiation dose received from the Chernobyl fallout was needed. The Frascati researcher constituted the control sample providing the information on how much background to subtract.

The screening of the three Italians, performed by scanning their bodies with a sodium iodide detector, showed clear peaks of potassium-40, a substance which all of us have in some measure in our bodies, as well as the more ominous signal due to the cesium-137 contamination. However, there was no large difference between the intensity of the cesium signal from the tested researchers and the control sample; the incident proved inconsequential.

Data!

The years 1986 and 1987 saw the completion of the CDF detector as it had been envisioned in the 1981 technical design report. Finally, hundreds of millions of collisions could be collected in 1988–1989, a data-taking period which came to be called "Run 0" only after 1992, when the following "Run 1" commenced. The acquired data raised huge interest due to the unprecedented collision energy (now 1.8 TeV). This allowed for the production of very energetic hadronic jets, as well as a significant number of W and Z boson decays. And many CDF collaborators were now confident that a top-quark signal would also eventually show up.

Despite the general optimism, the fruition of the data proved to be a long and grievous process. First of all, one needed to understand in detail the effect that the online trigger selections had on the acquired data, a task which by itself required several man-years of work. Then a precise reconstruction of particle trajectories had to be produced from the thousands of position measurements determined by the tracking chamber. The unprecedented number of charged particles produced by each high-energy Tevatron collision made this a novel challenge. The identification of hadronic jets and the measurement and calibration of their energy also required careful, detailed procedures that had to be designed from scratch. Furthermore, the definition of suitable recipes for the identification of electrons and muons called for a deep study of all their characteristics. Because of those difficulties, it took time before the experiment could start to produce significant physics results.

One of the first occasions for the CDF collaborators to prove their worth to the scientific community occurred in 1989, as they realized that they could achieve a "world's best" measurement of the Z boson mass, just as the Mark II experiment was about to do the same using the electron–positron collisions that the brand new SLC collider had started to provide. The competition between the Stanford Linear Accelerator Center (SLAC) and Fermilab was in those years quite fierce: both laboratories wanted to be recognized as the most important particle physics center of the United States. It was a matter of prestige as well as of access to research funds. In comparison, even the long-time competition between US laboratories and CERN remained in the background.

SLAC's director Burton Richter, who together with Samuel Ting had been awarded a Nobel Prize in 1976 for the J/ψ discovery, was one of the most notable advocates of electron–positron colliders, and he appeared critical of the hadron collider program of Fermilab. In 1988, Richter and Lederman (then at his last year as director of Fermilab) were both interviewed by Malcolm Browne for the *New York Times*. Browne's article, titled "*Search Quickens for Ultimate Particles*," featured Richter explaining how the SLC would soon measure the Z boson mass with a much better precision than what the UA1 and UA2 experiments had achieved a few years before at CERN. According to Richter, the large number of Z bosons produced by SLC would yield a deeper understanding on fundamental physics, answering questions on the existence of new generations of matter, on the Higgs boson, and a lot more.

Although the *Times* piece did not cast the Tevatron collider in a bad light, it did give the clear impression that SLAC was the real center of the action for US particle physics. That was already a bit annoying to CDF members. A still harder pill for them to swallow was hearing the random nasty comments that some SLAC scientists used to offer at conferences and in other public venues: they pictured hadron colliders as incapable of doing precision physics and as lacking sufficient resolution to measure the Z boson mass.

Of course, there was a bit of truth in those statements. The cleanest and most precise way to measure the mass of a resonance is not to compute it from the momenta of its decay products, but rather to perform a *formation experiment* where the energy of the collision is gradually stepped up in suitably spaced intervals in the vicinity of the mass of the particle. At each energy point, the rate of creation of the resonance is determined; the peak rate is achieved when the collision energy equals the particle mass. As the energy of electrons and positrons in the beam is very precisely known, the mass of the resonance can be measured with high accuracy. Such an *energy scan* may only be performed using point-like projectiles as electrons and positrons, and not with protons, as the latter are composite objects which yield an undetermined fraction of their kinetic energy to the collision. But the truth stopped there: even if hadron collisions could not scan the resonance shape, a well-designed detector could still measure the decay products of the studied particle with sufficient precision to produce a competitive measurement. Given a large enough number of observed decays, the issue was whether systematic uncertainties could be reduced as much as needed.

A crucial input on the feasibility of a precise Z mass measurement at CDF came in the late spring of 1989. Barry Wicklund, an Argonne physicist and one of the deepest thinkers in the collaboration, demonstrated during a talk at an Electroweak group meeting how a calibration of the energy measurement produced by the electromagnetic calorimeter could be achieved with a precision 10 times better than what had been until then thought possible. According to Brig Williams, Steve Hahn, and the other calorimeter guys, the electron energy calibration had to rely on test beam measurements. Data had been collected for that very purpose by directing a beam of 50-GeV electrons to the 15° wedges of the calorimeter before assembling them into the wheels that made up the central detector. The test beam data enabled a neat cross-calibration of the detector modules: this resulted in a 2% precision of the energy measurement. Such was a reasonable number for most measurements, but it was not nearly precise enough to allow a competitive measurement of the Z boson mass from the decay electrons.

Barry, who had worked at the problem with his Pennsylvania University colleague Fumihiko Ukegawa, demonstrated how it was possible to inter-calibrate the detector modules more effectively by using a sample of 17,000 electron candidates that he had isolated in the Run 0 dataset. Those were electrons produced by proton–antiproton collisions rather than by a dedicated test beam, but they were still a pure sample.

They came from well-reconstructed decays of heavy-flavored B or D hadrons, or conversions of energetic photons. Those 17,000 electrons were enough to study the so-called E/p ratio in each of the forty-eight 15° wedges that made up the two "wheels" of the central calorimeter. The E/p ratio is computed as the energy E of the electromagnetic shower initiated by the electron in the calorimeter, divided by the electron momentum p determined from the curvature of its trajectory in the tracking chamber. The ratio equals 1.0 for highly relativistic particles, but it is often measured to be a bit larger than that for electrons. Electrons emit photons in a magnetic field, so their momentum can be underestimated. The measured energy is instead unaffected by the photon emissions, since the photons land close to the electron shower in the calorimeter and their energy is automatically accounted for.

A careful modeling of the detector material in the tracker, whose amount determines the energy that electrons lose by photon radiation, allows one to connect the energy measurement to the momentum measurement using the E/p distribution. Thus, Wicklund had a way to calibrate the wedges of the electromagnetic calorimeter quite effectively, with uncertainties one order of magnitude smaller than those achievable with test beam data. Alvin Tollestrup, the spokesperson of CDF, was deeply impressed. After the meeting he approached Wicklund, insisting that the data-driven calibration be carried out in full: the E/p method could enable precision measurements of W and Z bosons.

"Why Don't You Do It?"

A few days later, the issue of Stanford's bad-mouthing of Fermilab became the topic of a lunch time discussion in the cafeteria of the Hirise. Tollestrup was sitting in the company of a dozen colleagues around the two big round tables they had set side by side. Steve Errede, an assistant professor from the University of Illinois who convened the Electroweak working group, was sitting across from Tollestrup. (In CDF, the progress of physics analyses was discussed within several independent working groups, led by pairs of conveners that remained in charge for two years; the Electroweak group managed analyses that targeted measurements of standard model properties in processes with electroweak bosons.) Everybody agreed that the

SLAC scientists had gone too far with their denigrating statements. One of them was reported publically referring to as "slimeballs" the CDF experimenters, making fun of their efforts to extract meaningful physics results from a collider unworthy of its fame.

The CDF scientists knew that from an objective standpoint the alleged imprecision of hadron collider experiments was baloney. For instance, they had already measured quite accurately the mass of the J/ψ meson using its decays into muon pairs. This showed that the tracking was capable of determining particle momenta excellently despite the "messy" hadronic environment and the large number of particles that proton–antiproton collisions generated in each interaction. And if SLAC scientists considered "precision physics" solely the measurement of the mass and properties of electroweak bosons, then again, that kind of precision physics was certainly also in CDF's cards. Using the couple of hundred Z decay candidates collected in the 1988–1989 run, a value much more precise than the one until then known appeared achievable, especially once Wicklund had showed how the electron energy could be calibrated to sub-percent accuracy.

One of the graduate students at the table said "We should just prove that the SLAC guys are wrong!", but that looked like wishful thinking. The members of CDF appeared resigned to the fact that in a very short time Mark II, the detector of SLC, would sweep the table with a Z mass measurement that from then on only the LEP collider at CERN would be capable of challenging. The first Z boson had been detected by Mark II on April 11, so a first measurement of the Z mass was considered imminent. At some point Tollestrup, who was sitting on the other side of the big table, looked in Errede's direction with his gorgeous, penetrating blue eyes, and said:

"So why don't you do it?"

Errede did not answer. He was taken by surprise; "Is he talking to me?" he thought. He had a great respect for Tollestrup as a scientist. In particular, he was deeply impressed by the transformation that Tollestrup had wrought on the construction of the superconducting magnets for the Tevatron after assuming the leadership of the project. Errede was also a bit intimidated by the fixed stare of those deep blue eyes and that snow-white thatch: "He was one of those guys, you wouldn't like to be in their gun sight," he recalls. And the Illinois professor could not be sure that Tollestrup was really singling him out from among all the colleagues sitting around the table; others could have been the target of the enquiry. A couple of seconds went by; then Tollestrup repeated his question insistently:

"Why don't you do it? Why don't you do it?"

Steve finally emerged from his momentary hypnotic state. Alvin was really asking him to direct the Electroweak group's forces toward a measurement of the Z mass. It had to be competitive with the one which Mark II was going to soon produce, and crucially it had to be published first. It was a fantastically tough challenge! But Steve knew he could count on quite a few strongly motivated colleagues to take it on. He quickly decided that there was only one possible answer, and he gave it.

"Yes, I think we can do it. Yeah, of course! Let's do it!"

Of course, being a fundamental parameter of the standard model, the Z boson mass did not need spite as a motivation to be measured. It was clearly in the interest of CDF to invest significant efforts to obtain a precise result. To Alvin and many others, the fact that Mark II was close to publishing their first Z mass measurement, whose foreseen precision CDF could not hope to attain, made the matter an urgent one. CDF had to produce a precise measurement before Mark II could publish its first result. Only thus would the work receive precious citations in all future physics papers discussing the Z boson. Otherwise the CDF members would be beaten by a nose, turning the small but precious bounty of Z bosons captured in Run 0 into a largely irrelevant dataset.

That same afternoon Steve assembled a group of willing and able collaborators, and identified and distributed the different required tasks to the crew. Then special weekly meetings were scheduled to closely monitor the progress of the activities. The ongoing activities were better not discussed at their otherwise natural place, the Electroweak group meeting: it was considered unsound to publicize too much the Z mass measurement effort within the collaboration. In particular, a few of the collaborators from the Berkeley group also participated in the Mark II experiment. They needed to be kept in the dark about the ongoing effort, as the possibility of leaks to the competitors had to be minimized.

The weekly cadence was soon sped up to two meetings per week, and by the beginning of July the participants ended up meeting multiple times every day, as it was realized that the analysis effort would otherwise not converge to a final result in time. The meetings were not a waste of time: there were many different analysis activities progressing steadily in parallel, and it was crucial that they could provide feedback to one another. First of all, there were two separate measurements: one based on the sample of 132 $Z{\to}\mu\mu$ decay candidates isolated in data collected by a trigger selecting events with high-momentum muons; and a second one using the 73 $Z{\to}ee$ candidates coming from data triggered by electrons of high energy. Those two measurements used the same calibration procedures for particle momenta and modeling of the physics of Z production and decay, and specific inputs related to the peculiarities of the measurement of electrons and muons. A multitude of subtle physics effects playing a role in the measurement required a detailed study and the full dedication of one or two scientists each. During those weeks, Steve felt like he was providing more psychological support than physics advice. He kept running up and down the trailer corridors from one office to the other to pass on information and provide specific input to his crew members. And he often found himself encouraging them and alleviating their discomfort for the previous commitments that they had been forced to set aside.

Unfathomable Code

The computer algorithm which performed the fit of position hits measured by the tracking detectors, extracting the most probable trajectory of charged particles, was still unfinished business in 1989. A working version of the code existed and did reasonably well, but due to the complexity of the problem, the program lent itself to further significant improvements. It was thanks to Peter Berge — the "tracking guru" of CDF, one of the many unsung heroes of the experiment — that the tracking algorithm reached the level of sophistication required for a precision measurement of the Z mass. Peter was considered a magician by his colleagues. Among other things, he was probably the only CDF member capable of effortlessly reading and writing in Postscript, a programming language used for creating vector graphics which is not meant to be written by humans, but by other programs. Berge worked alone and left quite rarely his small office in the oldest section of the CDF trailers complex. The office was often plagued with water leaks from the roof during rainfalls. He did not complain: when the dripping started, he used to place a bucket under the largest leak, close to the entrance of his office. He then taped up on the bucket a handwritten notice which read "Do not disturb — leak test in progress." The pun was for insiders only: a leak test is a standard procedure employed during the commissioning of a gas detector. It was a joking reference to the CTC, the focus of Berge's work.

Building on Berge's improved tracking code, Aseet Mukherjee was the guy who pulled off the remarkable feat of inserting the *beam constraint* in the fit of charged particle trajectories. The beam constraint was also Berge's idea. The tracking code in Run 0 did not use the knowledge of the beam position in the fit to charged particle trajectories. Neglecting that information made sense in general, as it meant treating in the same way particles that originated from the decay in flight of a long-lived parent and those that originated in the primary interaction. Only the latter have back-propagated trajectories that intercept the path of the proton and antiproton beams. However, it made a lot of sense to enforce that condition to the trajectories of the decay products of the extremely short-lived Z boson, using a mathematical constraint in the fitting procedure. Such a beam constraint would improve quite significantly the precision of the momentum measurement, since it amounted to adding to the track one very well-measured spatial point at one end of its helical path: a pivot.

In order to explain the procedure employed by Aseet to determine the coordinates of the collision point to be used for the beam constraint procedure, it is necessary to point out that the Tevatron collider bunched the circulating protons and antiprotons in 11-inch-long packets. Collisions at the center of the detector were consequently spread out along the beam by the same length. The position of each collision could be determined by fitting all the resulting charged tracks to their common origin. The precision obtained on the coordinate along the beam was more than sufficient, as its knowledge did not have a large impact on the fit of particle trajectories. On the other hand, the position in the plane perpendicular to the beam was crucial for the determination of transverse momenta. Using the knowledge of the coordinate along the beam and of the precise beam trajectory, which is a very thin line, Aseet could determine the transverse position of the collision with high precision. The only problem was that the procedure required quite a bit of fiddling with Terabytes of data, as hundreds of different fits on as many data-taking periods were needed to extract the required beam line parameters.

Aseet was well known for his mathematical skills. Once he produced the computer code that re-measured the trajectory of charged particles using the beam constraint, he was literally hailed as a hero by his colleagues. The better measured tracks had a dramatic effect on the momentum resolution, as demonstrated by the fact that the invariant mass peaks of particles reconstructed with the re-measured momenta narrowed down quite significantly. It was a proof that the measurement had gotten more precise. Encouraged by the general enthusiasm for his new method, Aseet kept changing his code on a daily basis for a while, providing a string of incremental improvements. That was a good thing, of course, but it forced everybody to re-run their own analysis programs one, two, three times a day in order to incorporate Aseet's improvements and benefit from them. It simply made people crazy.

Steve Errede decided to look into Aseet's source code to try and figure out whether the beam-constrained fitting could at last be signed off. That way, people would spend their time focusing on their own studies rather than on the spinning and re-spinning of the data. Time was running short, and the better looked increasingly an enemy of the good. Also, Steve was genuinely interested in how the algorithm worked and wanted to satiate his curiosity. When Aseet sent to Steve his fitting program, Steve suffered a shock. The code was long and complex, and this much he had anticipated. To make matters worse, the code did not contain any explanatory comments of the kind programmers insert at the most cryptic junctures amidst the Fortran instructions. That was already enough to ground Steve's attempts at making sense of the long computer routine. But, in addition, the names of variables and parameters were absolutely non-descriptive of their meaning. As a result, the program was almost as unreadable to a human eye as would have been its compiled image in the form of a machine-crunchable string of zeroes and ones. Steve had gotten more than he had bargained for! After spending a few sorry hours on the code, he decided that understanding it was not a priority; he had other things to worry about.

The kind of trouble that Errede ran into on that occasion was not an exception but rather a common occurrence in CDF. With few exceptions, particle physicists are not professional computer scientists. They usually acquire their programming skills by nose-diving into complex code projects when they join an experiment as undergraduates or graduate students. Hence, they typically learn to write very untidy, hard-to-read programs. They often completely neglect the good practice of inserting descriptive comments here and there. That is the result of being under constant time pressure. Also, young physicists are keen to use their own arbitrary conventions for the naming of variables and other coding choices. Unless the "Offline coordinator" (the scientist responsible for software development, who coordinates a group of software developers) enforces very strict rules and conventions for the programs that perform event reconstruction, calibrations, or other common applications, the result can be close to chaotic. However, in the case of Aseet's beam-constrained fit, the source of trouble were not his coding habits. Aseet was an expert programmer and he knew how to work under pressure. In order to quickly produce a version of the track fitter which added a fixed point to the track at the very precisely determined beam position, he had used a fairly complex mathematical method. The formulas had been taken from a scientific article, and Aseet had found it practical and time-effective to stick to the naming conventions used there, to have more control of what he was doing. The source of Steve's frustration was the choice of variable names in the article. Together with the code Steve had been pointed by Aseet to that reference; but he, in turn pressed by time, had not read it.

"Three MeV?! You Must Be Kidding Me!"

The CDF scientists involved in the Z mass analysis were all quite skilled and determined, and they worked around the clock for six weeks in a row.

Barry Wicklund painstakingly kept improving the analytical description of the detector material in the simulation of the apparatus. That gave a better agreement between the E/p distribution of electrons in data and simulation, and hence a smaller calibration error. Bob Wagner, an Argonne colleague of Barry's nicknamed "Argobob" to not confuse him with a homonymous colleague, wrote simulation programs to understand the effects of electromagnetic radiation in Z production, improving the modeling of the Z mass distribution. William Trischuk took care of generating large samples of simulated Z boson decays. Morris Binkley helped Aseet with the beam constraint code, which required as input the measurements of track positions in the vicinity of the beam. Those were provided by the VTPC detector, which Morris knew inside out. Hovhannes Keutelian, Errede's graduate student, developed and tested the program that produced a fit of the Z mass histograms: that was the code which would obtain the final result. The combined effort put together during those six weeks would pay great dividends in the future, since the problems the Z group had to solve stayed solved since then. The improved calibration of energy and momentum measurements, the more precise tracking algorithm, and the better tuned simulations would be an asset to CDF, reducing systematic uncertainties in hundreds of future measurements.

One of the crucial tasks was to calibrate the momentum measurement provided by the tracking chamber to as high a precision as possible. Only once that is done can the Z mass be precisely measured from the momenta of the two muons in Z→$\mu\mu$ decays; the calibration of the electron energy with the E/p method also requires that input. To calibrate the momentum measurement, one may use the tracks produced by the decays of J/ψ mesons to muon pairs. In 1989, the mass of the J/ψ meson was already known with an uncertainty smaller than 1 MeV thanks to earlier measurements, so its signal was a perfect "reference" for the calibration procedure.

Using the large sample of identified J/ψ decays to muon pairs, the measured J/ψ mass resulted 3 MeV lower than expected. This was a one-per mille downward bias. What was its source? The mass was measured with particle momenta, and momenta were determined from the curvature of their trajectories and the value of the magnetic field. Steve Errede and Bob Wagner could not believe that the precisely determined intensity of the magnetic field could be wrong by a part in a thousand. Maybe the bias was due to incorrectly modeled radiation effects? They decided to pose the question to the in-house theorist, Michelangelo Mangano.

Michelangelo was a young and brilliant theoretical physicist who in 1988 had joined the CDF experiment to work on QCD measurements. He had previously been a post-doctoral scientist at Fermilab in the theory division. There, he had developed new techniques for the calculation of strong interaction processes. This made him an invaluable resource for the experimental studies that the QCD group in CDF was starting to carry out. At that time, QCD was not yet a very well-understood theory, and the CDF measurements were extremely interesting. The Tevatron was stepping for the first time in a totally new energy regime, where the understanding of strong interactions needed to be tested with experimental data.

The basic processes which allowed Mangano to carry out his QCD studies were those yielding many hadronic jets, whose energy was back then also still in the need of a precise calibration. The typical uncertainties of measured jet energies were tens of GeV, hence thousands of times larger than the ones Errede and Wagner were puzzling over! When they explained their problem to Michelangelo, he could not help bursting into a hearty laugh: "Three MeV? M-e-V? Are you guys kidding me?" But indeed, a 3-MeV shift was a quite significant effect for the J/ψ. The statistical uncertainty on the mass measurement was just 1 MeV, so that 3-MeV shift could only be interpreted as a systematic bias. A failure to understand its source and correct for it would force Errede and Wagner to ascribe the bias to an unknown systematic uncertainty associated with the momentum measurement, significantly blowing up the uncertainty on the Z mass.

Michelangelo spent a good deal of time working at the problem from a theoretical perspective, but he could find nothing that could account for the mass shift. Indeed, final-state QED radiation did affect the measurement, but its effect was well understood and under control. The problem had to be elsewhere. And it was finally discovered that what was to be blamed was the experimentalists' habit of rounding numbers!

If you give a perfectly machined circular ring to an experimentalist and a theorist, asking them to tell you what is its circumference, the theorist will tell you it is π times its diameter and will stop there with this perfectly correct, albeit unspecific answer. The experimentalist will instead duly measure the diameter as well as she can with a ruler, then multiply that by 3.14, reporting the result with three digits of accuracy. She will not bother to use a dozen digits in the expression of π: the experimental uncertainty on the estimate of the diameter is larger than a few percent, so three significant digits for π are more than sufficient. However, if another experimentalist measured the diameter with a laser caliber, obtaining a value with six significant digits, and proceeded to multiply that by 3.14 forgetting that π is in fact 3.1415926..., she would be wasting the high precision of her apparatus! A similar thing was happening with the reconstruction code that spewed out particle momenta from fits to the track trajectories measured in the CTC. The conversion from curvature to momentum required an implicit multiplication by the speed of light, which equals 299,792,458 meters per second. Whoever had written the routine had not bothered to look up the exact value, inserting instead the approximate value of 300,000,000 meters per second commonly employed for back-of-the-envelope calculations. The difference was less than one per mille, but was still huge considering the high precision of the tracking measurement. That error had surfaced as a puzzling 0.1% shift in the momentum scale!

One PRL a Day ...

Toward mid-July, the CDF members finally realized they had in their hands a very competitive result. The Z mass was measured to be 90.9 GeV. The statistical uncertainty of the combined fit to di-electron and di-muon Z decay candidates produced by Keutelian was coming out in the ballpark of 300 MeV, and the dominant source of systematic uncertainty was the energy scale, amounting to 200 MeV. Overall, the CDF data allowed a fourfold reduction in the Z mass uncertainty over the current world average! It was time to wrap it up.

The article was written on the night of July 18, just hours before a scheduled *collaboration meeting*, a two-day event organized every three or four months where CDF members could listen to a detailed overview of the status of the experiment and the analyses that were being carried out. The publication was going to be a "Letter" — a short article to be sent to the prestigious *Physical Review Letters* (or "PRL" as everybody called it). As short as it needed to be in order to fit within the strict standards of PRL, writing a paper overnight was a challenging task for its editors. It was 6 AM on July 19 when the article was finally finished. Exhausted but happy, Steve went for breakfast, then drove to the Hirise. There the measurement could finally be presented to the collaboration. Sleep-deprived and physically drained, Steve had the task to convince his colleagues, who for the large majority had not had a chance to follow the progress of the measurement, that the result was solid enough to warrant an approval to be published without further scrutiny. The request of an approval without review was an unusual procedure. Normally, an internal review process was required before a publication could be sent out to a scientific journal. Fortunately, the response of the collaborators was very positive: this was indeed a careful, precise measurement deserving to be promptly sent to PRL.

As with any scientific publication, the title and abstract of the article were as important as the article itself and caused a lengthy discussion. On the title Errede managed to have it his way: "Measurement of the mass and width of the Z boson at the Fermilab TEVATRON." The capitalized name of the hadron collider was one of the points on which Errede stood his ground. The other was the explicit mention of the fact that CDF had measured the mass as well as the *natural width* of the resonance. The latter was something which electron–positron guys expected to be exclusively their own business, thanks to the already mentioned energy scan of the production rate.

The natural width of a particle can be determined by the distribution of observed mass values around the most frequent value. The width is inversely proportional to the particle's lifetime and is directly connected to the intensity of the interactions responsible for the particle decay. The stronger the interactions, the larger is the range of mass values that the particle may take, as the decay proceeds more quickly, giving no time to the particle to "settle" to its nominal mass. Arguably, the natural width of the Z boson is no less important than is the Z mass itself. Compared with model predictions, the measured width allows one to determine whether there may be decay modes of the particle that are not seen experimentally. A larger-than-predicted width implies the existence of unknown particles into which the Z can disintegrate, as mentioned in Chapter 1.

After the meeting Errede, still in a state of sleep deprivation from the past weeks and the frantic final night of work, met Alvin Tollestrup and Barry Wicklund. Together, they took the stairs to the second floor of the Hirise and went straight to John Peoples' office: they needed something from the Fermilab director. As instructed by Alvin, that morning Errede had sent to Peoples a copy of the article, along with a special request. The laboratory director was asked to undersign the paper before it got sent to PRL. The journal would ordinarily take several weeks to review an article. Suitable reviewers had to be found and be given sufficient time to comment on the results, require clarifications, propose modifications, or reject the article. The exchange between authors and PRL reviewers could easily push the publication date to September or October. Furthermore, the journal editors might choose a member of Mark II as reviewer, bringing in a conflict of interest. A SLAC reviewer might fall in the temptation of purposely slowing down the publication of the CDF paper. However, an article signed by the director of a major American physics laboratory allowed the peer review process to be waived. The article would go to press in the very next issue of the magazine.

The special no-review procedure had been instituted in 1976 by *Physical Review Letters* in what was called "An Editorial Experiment." The April 26 issue of the magazine featured a letter bearing that title, where the editors explained that the time to publication of even carefully crafted articles could be long, so they had cooked up a special procedure to bypass the regular review process:

> "To receive this special treatment, the Letter must be forwarded for publication by a division leader, department chairman, or director at the author's institution [...] This person, who may not be a co-author, must explain why special handling is appropriate. [...]. The editors reserve the right to evaluate the "extraordinary circumstances" and to refuse the request. [...] The published Letter will bear, immediately above the abstract, a note reading, "Published without review at the request of (person forwarding the Letter) under policy announced 26 April 1976.""

Peoples agreed to sign the paper and write a cover letter to PRL: it was certainly in the interest of the laboratory that CDF proved its worth in the competition with the Stanford experiment. Those were years when SLAC and Fermilab were competing on the amount of funding that they received from the Department of Energy. Richter had the reputation of being the one who won those Washington battles. The perception was that he was more persuasive and more charismatic than Peoples; the Nobel Prize he had won in 1976 also played a role. A joke in fact circulated among American experimental physicists on the matter: Richter and Peoples would both go to Washington to get money for their labs, and Peoples would be the one getting in the revolving doors at the entrance first, but the one who would get out of the doors first would always be Richter! However, the truth was different: Fermilab had actually a larger budget than SLAC at the time, but was starting to suffer from the ramping up of the funding to the Superconducting Supercollider (SSC). The Texas machine was a hadron collider, hence it fished in the same funding pond of the Tevatron.

Peoples did more than undersign the paper and the cover letter. That same day, he called the editor-in-chief of PRL David Lazarus, forewarning the arrival of the Z mass paper and asking whether the editor agreed to publish it in the next issue of the magazine. The editor denied the request, on the grounds that it would be difficult to organize the special handling of the article in such a short timescale. As he hung up, Peoples was left with an unpleasant suspicion. After discussing the matter further with his predecessor Lederman and with Tollestrup, Peoples called the editor again. Finally, the situation was clarified: Lazarus explained that SLAC had already pre-announced the submission of a Mark II paper on the Z mass, so he had set up an internal review panel to handle it. It was only through the stubborn insistence of Peoples that a similar panel was formed for the CDF paper.

"If You Show Me Yours, I'll Show You Mine"

While the frenzy at Fermilab was finally subsiding, an epic battle was brewing for Ken Ragan, a post-doctoral scientist at the University of Pennsylvania. Ken was at the time looking for a faculty position, and he needed to embellish his *curriculum vitae* with a talk at a major conference. A few months earlier he had discussed the matter with Brig Williams, the head of the Pennsylvania group in CDF. Brig had understood that the experiment would soon produce some important new result; perhaps he had surmised that a world's best Z mass was at reach. It was by following Brig's explicit suggestion that Ken applied to give the invited CDF overview talk at the Topical Conference following the 1989 SLAC Summer School.

By sheer chance, CDF was going to first present its new Z mass measurement in the lion's lair. Worse still for the Mark II collaborators, the conference organizers had scheduled the CDF talk on Wednesday, two days before the Mark II talk. This had the purpose of closing the conference with the announcement of the new SLAC results, and it also gave Mark II a few additional days for crossing all the t's and dotting all the i's of their own analysis of Z events. Fortunately, this arrangement also worked for CDF, as it gave Ken and his collaborators enough time to review the measurement and approve it. Before traveling to SLAC, Ken had purposely been instructed by the Electroweak group conveners to prepare two different talks. One included the Z mass measurement, and gave it the space it deserved. A second "backup" talk did not mention the Z at all and focused more on the other beautiful physics measurements that the experiment had produced. The backup talk would be the one to choose, in case some real concern about the soundness of the Z mass measurement arose during Errede's presentation at the collaboration meeting. But nothing like that happened.

On the morning of July 19, shortly before the start of the session, Ken was approached by a young Mark II physicist, whom I will call Arthur in the following. He cheered Ragan and soon made a strange request.

"Hey Ken, I hear that you guys in CDF have pulled off a measurement of the Z mass?!"

"Well, I don't know who told you that … But my talk is in two hours, so come and listen to it if you're interested!"

"Sure, I'm definitely coming. But why don't you tell me what is your result? You know, we also have a measurement … We are finalizing it and we'll send it out in just a few days. If you tell me yours, I can tell you ours."

"Hmmm, thanks, but I guess I'll pass. It doesn't matter much, but it would still be a violation of our internal rules … Just come to the talk!"

"Well, okay. It's fine. Look, here is our draft — just so you know that we do have it … You see?"

Arthur opened a folder and showed Ken the first page of a draft paper titled "Initial measurements of Z boson resonance parameters in e+e- collisions." The abstract showed the mass measurement obtained by Mark II, and Arthur made sure that Ken would read the number.

"I see ... Well, thanks Arthur. I need to get a seat now."

"Yeah, and good luck with your talk today!"

Despite Ken's refusal to share the CDF measurement, Arthur had reached his covert goal. Ken had read from the paper's abstract the value of Z mass that Mark II was measuring: 92.5 GeV, with a 0.2 GeV uncertainty. This was not incompatible with the old and less precise UA1 and UA2 results, which combined yielded 91.5 ± 1.7 GeV, but it was quite significantly higher than the result Ken was about to present! Was there something wrong with the CDF measurement? A less self-confident physicist would have decided to switch to the backup talk: after all, there were many other results to discuss in detail. But Ken did the right thing: he decided to ignore that extra bit of information. It was certainly strange and worrisome that the Mark II physicists would produce a measurement of the Z mass incompatible with the CDF one: their energy scan could not be affected by large systematic biases, so this implied a possible problem in the CDF measurement. Yet, CDF stood behind the result they had worked so hard to produce, and if their result were to be proved wrong, so be it.

Ken Ragan's talk was memorable. The measurement of the Z mass he showed was a little jewel. Many of his listeners did not know the first thing about the cunning methods that had been used to calibrate lepton energy and momentum. They listened in shock and awe. And the numerical result itself was shocking to the SLAC physicists. CDF had measured the mass with a total precision of 360 MeV, a fivefold reduction from the combined result of the two previous CERN experiments, a result competitive even with the systematics-free, lineshape-driven measurement that Mark II itself was about to produce! That was as close to a punch below the belt as an experimental physics result could be. It showed that hadron colliders could deliver competitive precision physics measurements despite the "dirty" collisions they were studying.

The intricacies of the Z measurement would have been enough for an hour-long seminar, but CDF had more to show. Using quite similar techniques, the W mass measurement was another world's best result. Further, measuring the two boson masses together allowed an important check of the standard model prediction for their ratio. The top-quark searches yielded an upper limit on its mass at 77 GeV, a result which again beat those that the CERN UA1 and UA2 experiments presented at the same conference. Furthermore, Ken's talk included a new, high-quality measurement of the cross-section of QCD processes and a first indication of the B-physics potential of the detector.

While the above events unfolded in the main auditorium at Stanford, Barry Wicklund was not in a jovial mood: Tollestrup had asked him to prepare a late afternoon "Wine and Cheese" seminar for the Fermilab physicists. Barry did not really want to give that seminar: it was a lot of work to put it together in the matter of one day. Barry was a power machine who would work like a truck when you gave him a difficult physics problem to solve, but he much less liked to spend his time on presentations. He had tried to redirect the task to Errede's student Keutelian, but Tollestrup would have none of that. Melissa Franklin finally convinced Barry by pointing out that he had been the source of the critical ingredient in the measurement, the electron energy scale, and that CDF wanted to acknowledge his contribution. So Barry spent the night of July 21 preparing his talk. When he finally drove to the Hirise he was caught by heavy rain and got drenched as he ran from the parking lot to the Ramsey auditorium. There Tollestrup introduced Barry as the star of the measurement team, but at that point Barry could not care less: soaked wet, he was feeling dead tired and demotivated. Yet he was an excellent speaker, and as he delivered his talk he did raise the interest of all the Fermilab audience. The lab theorists as well as the experimentalists from the other Fermilab experiments were all strongly impressed and almost in denial. Why, lepton collider specialists had been repeating to them that you could not do precision measurements at a hadron collider. And yet there it was, a 0.4% measurement of the Z mass, and a nice determination of the Z width to boot!

Aftermath

On August 9, Mark II released their paper as a laboratory preprint, numbered SLAC-5037, and sent it to PRL. The measurement consisted in a fit to the number of events collected as a function of the center-of-mass energy of the electron–positron collisions. This allowed Mark II to "see" the resonance shape, obtaining the Z mass as the location of the peak rate. The Z boson mass was determined to be 91.11 ± 0.23 GeV, a 35% reduction in uncertainty from the CDF result. But it had come out two weeks later! Scooped by CDF by two weeks, the Mark II collaborators had to content themselves with scooping LEP. The four CERN experiments would start collecting data five days later and soon produce results 10 times more accurate than those of Mark II. As for the CDF and the Mark II articles, they ended up in print in the same issue of PRL, which got published on August 14. Below the abstract, the Mark II paper reads "Received July 24th 1989." This is in mismatch with the preprint cited above, dated August 9. Regardless, one is left wondering why the CDF article had to appeal to the special 1976 rule of APS, whereas the Mark II one, which did not, still made it in the next issue of the magazine. Evidently, the Mark II paper benefited from the very same treatment of the CDF one — no external refereeing. The editor-in-chief Lazarus probably decided to avoid favoring Fermilab against SLAC. This was the correct decision; besides, rather than one against the other the two laboratories were really both competing against CERN. The LEP results would soon come in and send the American measurements to eternal oblivion.

91.11 ± 0.23 GeV is a very accurate measurement, as we know today that the Z weighs 91.19 GeV. But the alert reader will notice that the Mark II estimate lays a whole GeV and a half below the value that Arthur showed to Ragan before his talk! Did Mark II revise in a rush their measurement, finding a mistake at the last moment, after the events of July 19? That is extremely unlikely: a 1.5-GeV shift in the measured Z mass is an enormous one given the 0.23 GeV uncertainty quoted in the final Mark II article. But the alternative is almost as startling: it involves imagining that Arthur had purposely produced a fake draft with a bogus number in the abstract, in order to puzzle the CDF speaker — maybe in the attempt of convincing him to remove the Z measurement from his talk.

One may imagine that if, in place of Ken Ragan, CDF had sent to give the talk one of the slimeballs that some scientists at SLAC claimed populated in large supply the CDF author list, the outcome of the topical conference would have been a much sweeter one for Mark II. That fact alone was enough to raise some bad feelings in the CDF collaborators. Several of them received by electronic mail a summary of the events, where Ken included a mention of the pre-talk incident with Arthur. Worse still, a few Mark II collaborators reportedly let go with inflammatory statements. They openly complained to the Fermilab management, arguing that the escamotage of using the Fermilab director's signature to speed up the review process of their paper was an act of belligerence against SLAC. Some of them went as far as to speculate that CDF had used leaked information on the Z mass that Mark II was fitting, in order to ascertain that their own measurement was in the right ballpark! Of course, the SLAC scientists were either getting the story of Arthur's "leak" wrong or were being bad losers. The inflammatory accusation against CDF could also be read between the lines in an article by David Perlman on the *San Francisco Chronicle*, which was evidently embargoed by SLAC and got published on July 21:

> "What particularly annoyed the SLAC group, according to physicist Michael Riordan, is the fact that "Fermilab has very actively tried to scoop us by press release, even though their uncertainties are under serious challenge and they knew our measurements even before they released theirs." Reluctantly, however, Fermilab's Shochet conceded in a phone interview yesterday that Stanford's calculation was in fact predictably the more accurate."

This was only an inch short of accusing the CDF researches of scientific fraud, as it both questioned the uncertainty of their measurement and suggested that CDF might have used the knowledge of the "correct" Z mass — the one Mark II had estimated — to fine-tune their result! Many CDF members were shocked by the *Chronicle* article, and even more so by seeing that the Fermilab management did nothing to react to that absolute lack of fair play by the scientists of the other laboratory.

In retrospect, Arthur's trick and the allegations of the Mark II scientists leave a bad flavor to the whole story. Yet, it makes sense to try and discern a deeper meaning behind such petty behavior. Those reactions had their deepest roots in the strong belief that the SLAC scientists held about the impossibility of doing precision physics with hadron collisions. That was the gospel they had heard for years, as well as their motivation for planning a career in electron–positron machines and experiments. That gospel had brought them to sink in a groupthink black hole, one from which they were unable to get out. To them, a hadron collider could never be capable of reaching a 300-MeV precision on the mass of the Z, *because*! Because it was *obviously* impossible. Given this belief, their syllogism was immediate: if CDF had managed to produce such an accurate determination of the Z mass (they knew it was accurate since they measured the same value!), it meant CDF had cheated. The above syllogism explains the name-calling as well as Arthur's juvenile trick, yet the Mark II scientists were badly wrong: the hadron collider result was as genuine as theirs, and it was actually a groundbreaking scientific accomplishment.

As for the methods that allowed CDF to reduce the systematic uncertainties on electron energy scale, muon momentum scale, and all the other small nuisances that one has to worry about when performing sub-percent measurements, they have become cherished standards in hadron collisions and have enabled CDF to hold the world's best measurement on the W boson mass before and after the LEP II electron–positron collider, CERN's "W factory," got in that game. At the time of writing, the W boson mass is known with 0.02% precision thanks to the final CDF result, which is still the world's best.

Chapter 4

The Road to the Top

By the late 1980s, the main target of the CDF experiment had declaredly become the observation of the last quark still on the loose, the top. Yet, the road to the discovery of the top quark proved much longer and more difficult than anticipated, because of the unexpectedly large mass of that particle. The production of a heavy top quark was a quite rare outcome even for the very high energy of the Tevatron collisions; as a result, the collection of a significant number of signal events took longer than anticipated. In addition, the tools and the analysis strategies deployed by the early searches were inadequate, since they targeted a quark lighter than the W boson, which would display a phenomenology quite different from that of a heavier object. Those studies yielded a string of negative results. Hence, in spite of the great successes that the CDF experiment was reaping with the study of other physics processes, the general feeling was a mix of disappointment and frustration. The table was set and dinner was ready, but the VIP was not showing up!

Production and Decay of the Top Quark

In order to describe the distinguishing features of top-quark events, which can be exploited to collect data enriched of their signal, it is necessary to explain how top quarks are produced and what signatures they manifest when they decay.

At the Tevatron, the dominant top-quark production mechanism is the materialization of a top–antitop quark pair through the annihilation of a colliding quark–antiquark pair. We may imagine an up-type quark in the proton hitting head-on an antiup antiquark in the antiproton. The two particles turn into a gluon which immediately materializes a new quark–antiquark pair. A collision may thus transform an up–antiup pair into a top–antitop one. But there is a catch. In order for a top–antitop pair to be created, conservation of energy requires that the released energy be larger than twice the top mass, which today we know to be 173 GeV. In other words, the kinetic energy of the colliding up–antiup pair must be larger than 346 GeV, else the collision may only yield lighter quark pairs or gluon pairs. 346 GeV are a large fraction of the total energy theoretically available when a proton and an antiproton collide at the Tevatron (1800 GeV in Run 0 and Run 1). Further, the probability to find in the proton an up quark carrying a large fraction of the total energy of its parent is very small; and we need both the up and the antiup quarks to be energetic in order to produce a top pair. The process is consequently quite rare: it occurs only once in 10 billion Tevatron collisions!

As they materialize, the top and the antitop leave the interaction point in opposite directions. Yet, they do not go very far. The top quark lives a really short life: on average, less than one-millionth of one-billionth of one-billionth of a second. During that fantastically short time interval, it does not even manage to travel a distance equal to the size of a proton. And then it decays, becoming a bottom quark via the emission of a W boson.

W bosons also decay instantaneously. The carriers of "charged-current" weak interactions are very democratic in the way they interact with fermions: the strength of their coupling to leptons and quarks of any kind is identical. This is reflected in the proportion of their decays. The three possible decays of a W boson to lepton–neutrino pairs (namely $e\nu_e$, $\mu\nu_\mu$, and $\tau\nu_\tau$) are equally likely. The decays of a W to quark–antiquark pairs of the first or second family (up–antidown or charm–antistrange, respectively) instead occur three times as often as any of the leptonic decays. That is because quarks come in three possible colors, and quarks of each color "claim" a share of the W decays as each lepton family does. Note that W decays to third-family top–antibottom pairs are instead forbidden: the top mass is larger than the W mass, so such reactions would

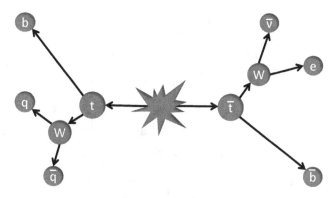

Figure 1. Schematic of the production and decay of a top–antitop quark pair. After their production in an energetic collision, top quarks undergo a decay to a W boson and a bottom quark. W bosons in turn decay into quark pairs (left) or into lepton–neutrino pairs (right). Antiparticles are identified by bars over their labels.

violate energy conservation. A schematic of the decay of a top–antitop pair is shown in Figure 1.

Taking stock, we can summarize the decay of a top–antitop pair as follows. Each top yields a W and a bottom quark. W bosons in turn yield a pair of quarks or leptons. Quarks then generate collimated streams of hadrons through the hadronization process described in Chapter 2. To be more quantitative, we may draw a box diagram where on each side we mark the relative frequency of decays of one of the two W bosons to each of its possible final states (Figure 2). There are a total of nine possible decays: three leptonic and six hadronic ones. Hence, each final state has a one-ninth probability of occurrence. As experimentally we cannot distinguish the six W decays to $q\bar{q}'$ pairs, which all produce pairs of hadronic jets, we lump them together as decays to pairs of jets; this final state results six-ninths of the time.

By counting how many squares correspond to the three differently shaded areas in the graph in Figure 2, and bearing in mind that there are a total of 81 squares, we obtain the different probabilities of top-pair decays. Thirty-six out of 81 of them generate events with six hadronic jets (two from the b-quarks and four from the W bosons), a topology called *all-hadronic decay*. Twenty-four more instead yield one charged electron

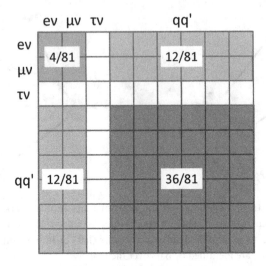

Figure 2.   Visualization of the relative probabilities of the possible final states resulting from the decay of a pair of W bosons. See the text for details.

or muon, a neutrino, and four jets — what is called *single-lepton decay*. Finally, only in four out of 81 cases the top pair yields the signature of two electrons or muons, two neutrinos, and two jets, which is dubbed *dilepton decay*. Note that final states with tau leptons (the 17 remaining white boxes in the graph) are excluded from this classification, since tau leptons are quite tricky to identify. In fact, in hadron collider physics jargon, the word "lepton" is often used (incorrectly but effectively) to indicate only electrons and muons, which provide very clean signatures.

I need to also mention that until 1991 physicists did not know that the top quark was heavier than the W boson, so they looked for events quite different from those belonging to the three above categories. If the top-quark mass is below 75 GeV, the particle can be created together with an antibottom quark in W decay; otherwise, the process is impossible, as already mentioned. This holds as long as the sum of the masses of top and bottom quarks is smaller than the mass of the W, which is 80.4 GeV. Because of the relatively large rate at which W bosons are created in Tevatron collisions, the production of a single top quark in W decay events was the most promising signature in the early searches.

Below I will indeed mention searches that assumed the top quark to be lighter than the W boson.

The First Searches for the Top Quark in CDF

The large number of collisions produced by the Tevatron and collected by the data acquisition system of CDF meant that a conspicuous amount of computing power was needed to reconstruct the observable particle trajectories from the electronic signals produced by the detector. A typical event could take several seconds to be reconstructed by a single processing unit. This was a real bottleneck for the analyzers, who were eager to extract physics results from the data. Yet, computing power was a scarce resource. Due to the high cost of commercial computer farms, in the 1980s Fermilab had decided to produce its own computers. The plan had initially been to supply the resulting computing power to its fixed-target experiments, which were already operational. The Fermilab machines were assembled using processors originally produced by Motorola for the first Macintosh computers. The processors were mounted in parallel on "VME boards," the hardware commonly used by experimentalists in the readout of their detectors. That arrangement was very cost-effective, but the use of the resulting computing power was by no means straightforward.

To use the custom-made computers, users wrote their programs on their desktop terminals using the VAX operating system, and *compiled* them there, producing executable code in machine language. (The compiler is a software that reduces human-readable instructions into sequences of binary code commands). The executable code would then be sent to the array of custom processing units, which served a batch queue. Unfortunately, the custom computers used a different compiler. That was the origin of a few small inconsistencies, as the language of the VAX executables was not 100% compatible with that of the Fermilab batch system. In addition, the custom processing units had very little memory. As a result, sometimes the programs sent to the batch queue would crash without warning: they stopped without producing a *core dump*, the special file which contains information on the cause of the crash and is usually generated when a program terminates abnormally. There was thus no easy way for the user to understand the source of the failure and patch it up. It was a software nightmare.

In the late 1980s, CDF had inherited the Fermilab batch system from the fixed-target experiments, and its researchers had a hard time processing the raw data coming from the detector: the reconstruction software was extremely complex, and computer crashes were the rule. The process of going from raw data to physics results was excruciatingly slow. This meant that possible new discoveries were being delayed! Fortunately, an *escamotage* was devised with the help of the University of Chicago. Bill Foster wrote from scratch a quick, two-dimensional reconstruction of charged tracks, which sped up quite significantly the slowest part of the event reconstruction. Then a dozen MicroVax workstations were connected together in the Chicago headquarters to create a "processor farm" to run the speed-effective reconstruction. There remained the problem of transferring the data to Chicago from Fermilab, which was 30 miles away! So a plan called *spin cycle* was implemented, which entailed the collaboration of a few young guys as transportation clerks. Claudio Campagnari, Paul Tipton, and Paul Derwent — the first two post-docs, the latter a graduate student — were "volunteered" to carry out that task. They took turns picking up the tape cartridges at Fermilab, still hot with raw data written by the CDF online output streams. They drove to Chicago, fed them to the processor farm, waited for the farm to produce its output, and brought back the processed data to Fermilab. This unofficial reconstruction took some shortcuts, but it allowed analyses to be carried out faster. Allan Clark, the Fermilab staff scientist in charge of the official offline reconstruction of the experiment, was not happy being thus bypassed. But he of course realized that the Chicago effort benefited the whole collaboration, so he did not interfere with it. The results that CDF was producing in 1988 all came from the spin cycle.

In December 1988, a one-day workshop was organized in the Ramsey auditorium, the conference hall in the basement of the Hirise. This workshop was specifically devoted to focused discussions on the top-quark search which several groups had independently undertaken. As I mentioned above, in 1988 the top quark was still assumed to have a mass lower than that of the W boson. Looking for a top quark much more massive than that, which would be produced at much smaller rates, was hopeless given the small amount of data collected until then. As usual, one looked for the lost car keys under the street lamp, since that was the only place lit up well enough to give some hope of success! So it was not events containing top–antitop quark pairs that were being sought, but rather ones containing a top–antibottom quark pair allegedly produced in W decays. The bottom quark would produce a hadronic jet, whereas the top quark would decay into a lepton, a neutrino, and a second bottom-quark jet. Because of the low mass of the top quark, the signature was therefore one of "lepton plus jets" events of smaller energy than those, similarly called "W+jets," which would allow the top discovery several years later (see Chapter 7).

The afternoon session of the workshop begun with a very detailed presentation by Jimmy Proudfoot and Barry Wicklund, who worked for the Argonne National Laboratory. The Argonne group in CDF was very active in the search for the top quark; the search used events containing an electron or a muon plus hadronic jets. They had tried to extract the top signal studying a variable called H_T, the sum of *transverse energy* (labeled E_T) of the jets plus the lepton transverse momentum and the missing transverse energy. This variable could discriminate the high-H_T top production process from low-H_T backgrounds. Jimmy and Barry had also looked in detail at many characteristics of the selected events. Their talk spurred a lively debate on the details of their selection and on the definition of the variables used to identify clean electron and muon events. In the end, everybody agreed with the conclusion of the Argonne researchers: there was no clear signal of a top quark in the data.

After the Argonne presentation, Pekka Sinervo got on stage. Working for the CDF group of Pennsylvania University, Pekka had looked for the top quark in a way similar to the one used by Argonne, and he was also reporting a negative result. Again his talk was received with wide interest, and the audience expressed criticism and direct suggestions. It was all in the right spirit of a workshop, where the best minds of the experiment collaborated to get the most out of the data they had worked so hard to collect. On the other hand, the picture was a bit depressing: the top quark was nowhere to be found. Furthermore, one got the feeling that a well-defined strategy for the top search was missing: people were throwing nets in the open sea, catching nothing. Claudio Campagnari, one of the post-docs of the Chicago group, was sitting in the back row and observed the proceedings. Tall and handsome, Claudio had dark hair, a big nose, and a clean-shaven face. He had joined CDF just six months before, but since then he had been too busy with the spin cycle to get involved in physics measurements. He had come to the workshop in the hope of learning what the status of the top search was. He wanted to start doing data analysis, and top-quark physics appeared an exciting topic to work on. That afternoon, however, his enthusiasm for the top hunt was being dampened by the bleak picture drawn by the presenters. This would change when Kunitaka Kondo took the podium.

In his early 50s, Kondo was lean and short, with black hair and lively black eyes; he wore thick square glasses and usually dressed in elegant black or grey suits. A charming and very polite person, the perpetually smiling Kondo looked like nothing could ever upset him. Kondo's gentlemanly manner threw his collaborators, tuned to a much rougher and no-nonsense style of interaction with colleagues, totally out of whack: they did not know how to handle him. I recall that a few years later, as I joined CDF as an undergraduate student, he treated me with the same courtesy and respect he used with everybody else, from janitors to full professors. Whenever I crossed him in the corridors of the CDF trailers, he would stop, smile, and slightly bend his head and shoulders forward in a respectful nod. At that point, I could not help clumsily doing the same. I would invariably spend the following five minutes wondering whether my bending angle had been sufficient, just right, or excessive.

After discussing his participation to the experiment with Alvin Tollestrup, who had famously suggested that he bring "five people and five million dollars" to strengthen the collaborative effort, Kondo in the early 80s had led a group of 15 Japanese scientists to join CDF, which back then still counted just 87 collaborators. That 15% fraction would roughly remain the same throughout the lifetime of the experiment, growing in synchrony with the collaboration size. The Japanese team had contributed in many ways to the successful commissioning of the detector: the completion of the design and construction of the central solenoid, the electromagnetic calorimeter, and the tracking and muon chambers. Yet, Kondo's passion and focus was data analysis. The Japanese scientist had devised a very complex, cunning method to discriminate top-quark events from background ones, based on an analysis approach he had dubbed *dynamical likelihood*. His method would become a sophisticated and appreciated tool a decade later, but was received with wholesale skepticism at the time.

Kondo's dynamical likelihood method consisted in constructing probability distributions for the observed kinematics of the jets and leptons in the events. The distributions were used to derive a mathematical function describing the likelihood that the events were more signal-like or background-like. The likelihood could be finally used as a discriminant variable to select signal-enriched samples of data or to directly extract signal properties. It is ironic to observe that nowadays the most precise measurements of the mass of the top quark rely on the method called "matrix element," which is nothing but Kondo's original idea recast in the context of a measurement of the top mass rather than the discrimination of a top signal. Kondo was way ahead of his time, and like most pioneers in science he struggled to get his work accepted in an environment dominated by a conservative groupthink. In 1998 he would be awarded the prestigious Nishina Prize for his contributions to the top-quark discovery; yet 10 years earlier his ideas were not yet given the attention they deserved.

The problem in truth did not only lay in the boldness of Kondo's advanced statistical techniques. The Japanese physicist was also a bit clueless regarding many of the subtleties required to perform data analysis in the experiment. His studies would contain silly omissions, such as neglecting the effect of the trigger selection, which modified the distributions of observable quantities; or not using the group-approved recipes in the selection of good lepton candidates. Those were minor faults, which could have been easily corrected. They by and large had their root in the lack of communication between Kondo's group and the rest of the collaboration. Unfortunately, his unfathomable techniques and the omission of simple but important details had the effect of discrediting his results. His colleagues in CDF felt it would cost them too much time to get Kondo's basic selection and analysis tools aligned with what was considered a standard. Nobody had the patience to even tell him what was wrong. On the other hand, checking his work by repeating his analysis from scratch appeared impossible in view of the lack of a full documentation of his novel algorithms. Those looked like inscrutable black boxes, which could only be mastered by their original creator. Slowly, Kondo came to be considered a lost cause. His politeness and good manners differed so much from American standards that they ended up creating a barrier between him and his collaborators.

It is by now four in the afternoon, and Kondo finally gives a full status report of his analysis in the Ramsey auditorium. The presentation is thorough and yet almost unintelligible by most of his listeners. Besides the use of his dynamical likelihood, the analysis employs brilliant but highly unorthodox tricks, like taking a jet from a simulated event and plugging it into a real one to verify the working of his algorithm on more complex events. His colleagues listen in an atmosphere of disbelief mixed with awe; some of them cannot hide a malicious grin. Nobody interrupts the speaker, despite the complexity of the material and the possibility to object on a hundred of details, or to ask for direly needed clarifications. As Kondo reaches the end of his talk, he concludes with a volume of voice just one decibel higher than the rest of his speech:

"And therefore," a pause, and then "I think we have discovered the top quark!"

The audience remains silent. Campagnari suddenly pulls up from the laid-back 150° angle that his legs and torso have ended up making as he sat in the back-row seat, making it an "am-all-ears" 80° one: what? Where? When? The workshop convener is Brig Williams, a professor from Pennsylvania University. He is a tall, lean guy with a sharp nose and a penetrating stare; he looks like an English gentleman from a 19th-century novel, especially thanks to his considerable aplomb. He is not impressed by Kondo's claim in the least.

"Thank you very much, Kuni. Is there any question?" One, two, three, four, "… No questions. Okay, thanks again Kuni. We stop here for a coffee break, let's be back at 5 pm sharp."

Brig Williams' apparent disregard of the work of an esteemed foreign colleague might sound rude. Still, in those years CDF was not a place where people would exchange courtesies and compliments — alas, it never was one. The only way to earn the respect of colleagues was through your hard work and your scientific merits. Conversely, if your analysis methods were not deemed publishable, or if your results were thought fallacious and your claims unsupported, you would be considered a potential threat to the good name of the experiment, and would suffer little short of boy-cott. In retrospect, the way Kondo was treated was gentle and friendly, in comparison to the rough ride that was reserved to other eclectic collaborators along the way to the top discovery.

One Event Too Many

The early searches clarified that the top quark was nowhere to be found in W boson decays. It soon started to dawn on the CDF top seekers that the elusive particle had to be heavier than the W boson. Hence, the focus moved to the production of top–antitop pairs via the strong interaction, a process which guaranteed more distinguishable signatures at the cost of lower rates.

The search of dilepton decays was the most promising way to identify top–antitop pairs in the early searches. There are very few competing processes yielding the signature of two energetic electrons or muons. Or even better, an electron and a muon, since that combination does not suffer from the background coming from the production of a Z boson with subsequent decay to an electron–positron or muon–antimuon pair. When the two leptons are accompanied by a large amount of missing transverse energy and two hadronic jets, the decay of a top pair is by far their most likely origin. The problem with this decay mode is its rarity, as it demands that both W bosons created in the top–antitop decay yield electron–neutrino or muon–neutrino pairs. Those combinations occur in a total of only 5% of the decays (see again Figure 2). Knowing as we do today that the top-pair production cross section in Tevatron collisions is about 5 picobarns, one may conjecture that just one dilepton event was produced in the four inverse picobarns of collisions collected during Run 0.

The calculation of the above prediction is simple, and it is worth repeating it here. The number of dilepton events is obtained by the product of three numbers: σ times L times B, where σ is the top-pair production cross section, L the integrated luminosity, and B the *branching ratio*, the probability of the considered decay mode — 5%, as stated above. And 5% of 5 picobarns times four inverse picobarns equals 1! One expected event. But this is before one considers that electrons, muons, and jets all must fulfill some kinematic and quality requirements in order to be detected and correctly identified. Those conditions unavoidably reject at least half of the produced signal. Finding a clean dilepton event in the collected Run 0 data can thus be considered a lucky chance.

One very striking event of that kind was indeed found. The event had observed characteristics closely matching those expected from the decay of a heavy top pair: a high-transverse momentum electron, a high-transverse-momentum muon, and one additional hadronic jet. It was promising, even exciting; but not a discovery-level observation, not even just evidence of anything. In the 1990 article describing the CDF top search, the authors noted:

> "With leptons at such high transverse momenta, the interpretation of this event as background [...] is unlikely. [...] A firm conclusion about the identity of this event is not possible."

The event could in fact have been due to mundane processes not involving top quarks. One of those was the production of a Z boson decaying to a pair of tau leptons: the decay of the two tauons could yield the observed electron and muon. The small rate of such a process made it an unlikely explanation, yet it could not be ruled out. Perhaps, two or three events of similar characteristics would have sufficed to put forth a tentative claim of seeing top production, but that singleton certainly proved not enough to convince the collaboration.

To make matters worse, in a dilepton event the top mass is not directly measurable from the energy of the detected objects. This is at variance with the case when all the bodies produced in the decay are identified and their energy and direction measured, when producing an estimate of the mass of the decayed particle is straightforward. In dilepton decays, each top quark produces one energetic neutrino through the decay of a W boson. The two neutrinos escape undetected, hence their direction and energy are unknown. That is a pity, as the knowledge of the top mass allows a theoretical estimate of the production cross section, yielding in turn a prediction of the number of signal events that should be seen. This enables a numerical comparison of observed and predicted events, which is a good verification of the signal interpretation. A mass measurement also enables more precise tests and targeted searches in the other decay channels.

That dilepton singleton was at least one event too few. Yet, from a different standpoint, it also proved to be one too many. Krzystof Sliwa, a CDF member from Tufts University, had devised together with his colleagues (and non-CDF members) Gary Goldstein and Richard Dalitz a method by means of which they could indeed determine, albeit indirectly, the most likely value of the top-quark mass from the apparently insufficient measured quantities in dilepton events. During the spring of 1992, Goldstein was invited to present and discuss his method at several meetings of the Heavy Flavor group of CDF, but no conclusion was reached on the soundness of the calculation. The Goldstein-Dalitz-Sliwa method did not convince CDF. Besides, the group members did not like the idea that outsiders could fiddle with their precious data, although the characteristics of the dilepton event candidate had been public since their 1990 article. The general feeling was that the event could not be used for any definite claim. To the group members, it looked much more principled to wait for the new data that the impending Run 1 was about to provide in larger amounts.

Despite the negative feedback, Sliwa appeared convinced that the isolated event had all the required characteristics to be a genuine top-pair decay, and that the top-quark mass was in the ballpark of 135–140 GeV as suggested by the mass reconstruction of Dalitz and Goldstein. That estimate is recognized today as way off the true value, but at the time it appeared credible, as it roughly matched with the indirect information provided by the data that the four LEP experiments had started to produce from 1990 onward. In pursuit of a proof that the dilepton candidate constituted solid evidence of the top quark existence, Sliwa studied with Dalitz and Goldstein how their mass reconstruction could be adapted to the few single-lepton top-pair decay candidates that CDF had isolated in Run 0 data. That was a logical step to take; but it created a problem to the collaboration as soon as the three physicists documented their ongoing work in a couple of *CDF notes*, the internal documents used by researchers to distribute information within the collaboration.

The documents discussed in detail the Run 0 single-lepton top candidates. They were taken as a written proof that Sliwa had distributed internal CDF information to non-collaboration members. As opposed to the above-mentioned dilepton event, the single-lepton candidates had not been published by CDF in all their characteristics and were thus considered "internal information." CDF note 1751, dated May 15, 1992, tried to demonstrate that the reconstructed mass distribution of the selected events was incompatible with the W+jets background and that it suggested a top-quark mass in the 120–140 GeV range. In the last sentence of the document, the three scientists quoted the original article by Dalitz and Goldstein on the dilepton event and the mass they had determined there, implicitly pointing out the compatibility of the two estimates. Although legitimate and not in violation of CDF rules, that publication had already been considered a form of encroachment by some CDF members. But the sharing of the single-lepton data suggested by the documented analysis was a clear violation of the rules of the collaboration. Sliwa denied having distributed the event characteristics to Goldstein and Dalitz, claiming that he alone had worked at the application of the mass reconstruction algorithm to CDF data. Yet, the names of non-CDF members on the internal document allegedly constituted a breach of CDF bylaws.

After internal consultations and discussions at stormy Executive Board meetings, Sliwa and his Tufts collaborators suffered some censorship, but the matter ended there. As for the mass reconstruction method of Sliwa, Dalitz, and Goldstein, it was long afterwards accepted as a sound one. With few modifications, it ended up being used in several measurements of the top-quark mass from dilepton events. In similarity to what happened with the dynamical likelihood of Kunitaka Kondo, the advanced method proposed by Sliwa was ahead of its time. The lesson we can draw from the two episodes is that if you have a brilliant new idea on how to analyze the data, you have better try and convince your peers of its value long before you attempt to use it to produce some bold new claim!

In the end, neither the lone dilepton candidate nor the few single-lepton events could provide a clear indication of top-quark production. From the absence of an observed signal, the experiment set a lower limit at 91 GeV on the top-quark mass: a lower-mass top would have been produced with too large a rate to be missed. That result was the final word of

CDF on the top quark from data collected in Run 0. It was a disappointing outcome, yet it left every hope alive for the coming new run, which promised to deliver much larger statistics. And CDF was now deploying a new, powerful weapon for the top-quark search, one which would tremendously increase the discovery potential of the experiment: the silicon microvertex detector (SVX).

A Heart of Silicon

Silicon detectors were an emergent reality in the 1980s. The concept of high-precision detection of charged particle trajectories using a solid-state tracking device was a recent revolutionary advancement. Until then, silicon had been used mostly in electron–positron collisions; in hadron machines its use had been limited to fixed-target experiments. There, layers of silicon sensors were employed as an active target wherein one could detect the short decay length of charmed particles and tau leptons. The design was one commonly used in neutrino beam experiments: by packing many layers of sensitive sheets into a stack positioned orthogonally with respect to the direction of incident particles, the trajectories of charged particles could be precisely tracked using the signal they left in each sheet.

Silicon detectors may be thought of as the solid-state analogue of wire chambers (see Chapter 2), as both devices record the passage of charged particles by collecting ionization charge on electric components. In silicon detectors, silicon crystals take the place of the gas of wire chambers; however, in the crystals no multiplication of the ionization signal takes place. The crystals are cut in the form of rectangular 300-μm-thick tiles, and they are *doped* with different atoms on the two sides. The atoms of arsenic and boron, once incorporated in the lattice of silicon atoms, respectively yield one electron to the lattice (donors), or steal one from it (acceptors). Two layers are formed, one with an excess of electrons, and another with an excess of "holes" — atoms with one less electron than they would need to be neutral. This so-called silicon wafer allows for electrical current to flow freely from one side to the other of the 300-μm junction, if a potential difference is applied across it in the correct direction (the one such that the hole-rich side has a positive potential attracting the electrons).

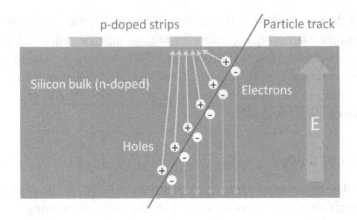

Figure 3.   Cross-section view of a silicon wafer. A voltage difference is applied to the two sides of the crystal, creating an electric field across the junction (E, upward pointing arrow on the right). The sensor is composed of an arsenic-doped (n-doped) bulk with thin boron-doped (p-doped) strips (here shown in cross-section). The strips collect an ionization signal produced by the crossing of a charged particle.

Practically no current may instead flow in the opposite direction. The system behaves as a *diode* junction (see Figure 3).

If one creates across the junction a potential difference of the order of 100 V in the direction opposite to that which would allow charge to flow, the silicon wafer behaves as an insulator, as the potential difference fully depletes it of free charge carriers. Electrodes on the positive side, shaped in very long and thin microstrips spaced by just a few tens of microns, may then collect the signal of the about 20,000 electron–ion pairs that are typically produced when a charged particle traverses the silicon layer. The narrow pitch of the microstrips guarantees an extremely precise position measurement orthogonally to the strips direction. Therefore, if one is interested in measuring the curvature of tracks in a magnetic field, one will arrange many layers of sensors along the particle path, in such a way that their strips are orthogonal to the plane of curvature and thus parallel to the magnetic field. This is the natural choice at modern collider experiments, where the magnetic field is solenoidal and aligned with the beam direction.

As explained in Chapter 2, in the design of a collider experiment, the order with which one collects information on the particles produced in the hard collision is of special importance. Until the advent of silicon detectors there was no alternative to low-density gaseous detectors for the first device intercepting charged particles, which was meant to measure particle trajectories without appreciably modifying their momenta. Only downstream of a successful track measurement could the particles be made to deposit all their energy through their destructive interaction with the high-density material of a calorimeter system. Because of that undisputed paradigm — track first, destroy later — tracking charged particles in a solid material was anathema: the atoms of the material would increase the chance of nuclear interactions and scatterings, worsening the successive energy measurement as well as creating additional debris around the particle track. Solid-state trackers could still be considered for very clean, low-multiplicity and low-rate collisions produced by an electron–positron machine. At a hadron collider, which generated hundreds of particles in every interaction at a much higher rate, their use was unthinkable.

In addition to the problem presented by nuclear interactions, the use of silicon microstrip detectors in a hadronic machine involved several challenging technical issues. The reading out of the tiny electric charge signals deposited in the thousands of strips was a slow process: it appeared impossible to adapt it to the very large interaction rate of hadron collisions. And hadron colliders also presented another challenge, due to a combination of design factors and electronic specifications of the silicon sensors. The interaction region at the Tevatron was spread along the beam line in a region about 12-inches long. Silicon layers built around it needed to be at least twice as long if they were to intercept most of the particles created in the collisions and to have microstrips running longitudinally, as explained above. Instrumenting a 2-ft-long detector with proper readout and power cables while keeping the amount of passive material at a minimum called for the use of long strips. But longer strips have worse electronic properties. This would cause a large noise, which made the extraction of a signal very hard to obtain.

In the light of those technical issues, it was no surprise that initially only a tiny minority of the CDF members believed that a silicon microvertex detector could improve the experiment's capabilities. The old 1981 technical design report of the experiment did contain a mention of the possibility of inserting within the CTC a small vertex detector made of four layers of silicon strips. The document also included a sketch of how those layers could be arranged around the beam pipe. But that was maybe more wishful thinking than long-term vision, because the technology to implement such a plan did not yet exist.

Five years after the 1981 report, a crucial step in the planning of the SVX was taken. A meeting called by Alvin Tollestrup, who was then co-spokesperson of the experiment with Roy Schwitters, was held at the Hirise with the aim of discussing the possibility to endow the center of the detector with a precise vertex-finding device which could complement the CTC. Besides the skepticism on the use of silicon detectors in hadronic environments, high-energy physicists in 1986 had clear in their mind the dangers of instrumenting with dense materials the inner region of a collider experiment. The input was coming from the trouble recently faced by Carlo Rubbia's UA1 experiment at CERN. UA1 had a small wire chamber in its core, installed within the main drift chamber. The collaboration had to face big problems because of the particular geometry of that detector, as particles emitted at small angle from the beam direction frequently hit the chambers' end plates, generating showers of secondary hadrons. Bent by the magnetic dipole field, the hadrons often managed to enter the main drift chamber, creating a large background which proved very hard to suppress.

At the Hirise meeting, Aldo Menzione showed a transparency where he had sketched with colored markers the construction schedule of the SVX as he envisioned it. The idea was to start working at it on the day after the meeting; the tight schedule was supposed to lead to the installation of a commissioned detector in the heart of CDF in a matter of two years. This appeared wildly unrealistic to Aldo's audience. Yet in hindsight, Aldo's vision must be acknowledged: he trusted that the silicon detector could work despite all its technical challenges, and he felt that it could help CDF in the top-quark discovery. He was right, and he had a deserved last laugh in 2009, when he shared with his Pisa colleague Luciano Ristori the prestigious Panofsky prize

> "for their leading role in the establishment and use of precision silicon tracking detectors at hadron colliders, enabling broad advances in knowledge of the top quark, b-hadrons, and charm-hadrons."

At that 1986 meeting, attendees were confronted with the tough challenges implied by the new technology, compounded with a dubious physics case. At that time the top quark was still believed to be lighter than the W boson, so its discovery did not call for the improved identification capabilities that a silicon detector might offer, even assuming that it could work as envisioned. Furthermore, the potential of a silicon microvertex detector for the physics of B hadrons (particles containing b-quarks, whose study critically relies on precise tracking) was totally beyond everybody's vision. Nobody expected that the Tevatron would end up producing the high integrated luminosity it eventually yielded. Nobody even dreamt that the B hadron physics studies enabled by the SVX and strengthened by large datasets would bring CDF to compete in that field of research with LEP and dedicated B hadron factories. In 1986, silicon looked like a gamble not worth taking.

Despite the general skepticism, the Berkeley physicists present at the meeting were intrigued by the discussion. At their institution there was interest in developing the silicon technology for the detector of the superconducting supercollider (SSC), the 40-TeV accelerator planned to be constructed in Texas. So the Berkeley physicists started looking into the problem of how to read out the silicon strips faster than had until then been accomplished.

Carl Haber, a young member of the Berkeley group, soon took the lead of the design of *front-end* electronics, the hardware responsible for a first handling of the detector output. His first task was to study a microchip that had been developed at Stanford, called the *Microplex*. He got a few samples of that chip and built a test stand to study its performance. He soon understood that it was indeed too slow and too noisy to be useful for CDF, and it used too much power. So he looked for help, and he got lucky. At the Lawrence Berkeley National Laboratories, a young electrical engineer called Stewart Kleinfelder had trained himself to do amazing things with electronic chips. Haber asked Kleinfelder whether he could make something faster than the Microplex, and with better signal amplification properties. Kleinfelder showed his genius as he came back in no time with a drawing on a napkin. The sketch showed a circuit that did precisely what was needed. The circuit had pre-loaded charge thresholds for the channels to be read out: if a channel did not pass that threshold, the electronic circuit would ignore it. He had invented the *sparsified readout*: by reading the silicon strips in this sparsified mode, the data acquisition speed could be improved by orders of magnitude! Nobody had managed to do that before, as it was believed that the charge in channels not hit by particles varied too much to allow the use of fixed thresholds. If the system could be made stable enough, Kleinfelder's idea could work. The technician had a few test samples of the chip produced and he designed a circuit to verify the sparsified readout. This showed that the idea would work. The way to circumvent the major obstacle in using silicon detectors at hadron colliders had been found.

In the meantime, Aldo Menzione collected a group of researchers from the University of Pisa to design the mechanical structure and a proper cooling system for the would-be detector. Even with the low-power-consuming sparsified readout, the front-end electronics would generate a lot of heat. Unless one provided a means to expel the excess heat, the overhearing would result in the electrical and mechanical failure of the detector. The goal was to show how one could construct a detector weighing less than 1.5 kg. That was in order to keep at a minimum the nuclear interactions of hadrons produced in the collisions. The Pisa scientists found how they could mount the silicon layers on a rigid yet light-weight beryllium structure, and thin aluminum tubes flowing cold water were shown appropriate to cool and thermally stabilize the detector.

Slowly, the results of the Berkeley and Pisa studies convinced Tollestrup that a silicon detector could actually be constructed, and he gave a green light to write a technical proposal for the device. A group was soon formed by all the physicists interested in contributing to the project. Aldo Menzione and his colleague Franco Bedeschi started to recruit personnel among the young post-docs in Pisa; along with them, the CDF group in Berkeley led by Bill Carithers took part in forces; there, Haber took charge of the development of the electronics. At Fermilab, Dante Amidei was designated to lead a group of a few scientists. That was it — the construction of the most innovative piece of the CDF detector was started by a dozen scientists, initially surrounded by the general skepticism of the rest of the collaboration.

Besides the understandable skepticism on the technical feasibility of the detector and its operation in a hadronic environment, the question was also what to use the SVX for. Menzione's original idea, conceived when the top quark was still believed to be lighter than the W boson, was to distinguish the top from the bottom quark by looking at the precisely reconstructed trajectory of the electrons and muons emitted in the decay.

When the lepton is created in the practically instantaneous decay of a top quark, it is emitted directly from the interaction point. When instead the lepton comes from a long-lived bottom quark, it exhibits a back-propagated trajectory that does not intersect the interaction point. The SVX information was thus meant to be used to *veto* the background from B hadron decays in the definition of top candidate events. It was only after 1990, when top-quark searches started to focus on top quarks heavier than the W boson, that the idea of actively identifying b-quark jets took shape. Hence, Menzione's original plan of a b-quark veto, while paradoxical in hindsight, played an important role in justifying the addition of a silicon detector to the CDF tracking system.

Loose Cannon

In February 1990 Nicola Turini, a young Ph.D. student from Pisa, arrived at Fermilab to test the first silicon wafers. The wafers were assembled in a two-layer test arrangement and were provided with the readout chip produced by Berkeley. A clean room had been set up in IB4, an industrial building across the road from B0. At the beginning, Nicola was the only physicist working full time on the silicon sensors. Carl Haber would spend every other week there for some periods, or come on Friday from Berkeley, work day and night during the weekend, meet on the following Monday at 8 AM with the other team members for a short status report, and fly back. The leaders of the Pisa group would only make extemporaneous visits.

The silicon sensors were delicate instruments, and a clean, dry environment was necessary in order to avoid damaging them during their handling. Nicola once got quite distressed when Aldo Menzione, who had arrived for a short trip from Pisa and was visiting the facility to supervise the activities, grabbed one silicon wafer with his big hairy hand and sprayed its surface with alcohol, then cleaned it by wiping the liquid off with a piece of cotton. "I want to see if it is really so delicate" was his explanation when Nicola tried to object. It was, and there went one of the first two working sensors! But luckily, soon other detector components arrived from Micron, the English factory that had agreed to produce the batch of sensors necessary to instrument the SVX. The new wafers were found to be working well: that was reassuring, since the technology of silicon wafers was ramping up in those years, and their production was still not as streamlined and faultless as it would later become.

Soon Nicola was joined by his Pisa colleagues Mosé Mariotti and Francesco Tartarelli. Other physicists from Pisa and Berkeley occasionally worked with the three Italians at the project, but the group of hardware developers who spent their time testing the silicon wafers and the readout system at Fermilab was comprised of three 26-year-old boys with quick minds and some dexterity with electronics. In earnest, it did not look like the team one would rely upon in order to test the most delicate components of the whole CDF detector. But they worked with dedication and did not spend too much time informing the leaders of the project on the actions they took. Amidei had named Turini "Loose Cannon" because Turini could not be directed and was thus capable of doing a lot of damage. He would do whatever he saw fit, regardless of the indications he received during the Monday meetings. But it paid off.

The problems the team faced were apparently insoluble. The detectors could not be read out the way that had been imagined when the readout chip had been developed. A silicon wafer contains hundreds of electronics channels. Each channel can produce a signal in the form of an impulse of current, created by the charge deposition of an ionizing particle on the silicon strip. But the strips also continuously yield spurious "leakage current" in the absence of excess ionization. To avoid reading useless channels, the chip developed by Kleinfelder at Berkeley allowed the sparsified readout mode as an option. Yet, every time Nicola and Mosé tried to use that feature, the output data became impossible to understand due to electrical oscillations.

In order to test the four-layer system that they had assembled — a *silicon telescope* built as a stack of four wafers — the young SVX crew needed a beam of particles of known characteristics. So, they spent their time in the Meson area of Fermilab, where the endcap calorimeter of CDF had been placed to collect *test beam* data. A test beam is what it sounds: a beam of specific characteristics which is used to test a detector or other device. The beam was the result of the interaction of 120-GeV protons from the Fermilab main ring with a fixed target. Particles of the required species and momentum were selected by the operators in the main control room using a system of magnets downstream of the collisions. The SVX team had been allowed to use the beam that the calorimeter was receiving, thanks to an agreement with the leader of that project, the young Harvard professor Melissa Franklin.

A box containing the silicon layers was placed few meters upstream of the calorimeter module under test, such that the beam impinged on it before reaching the calorimeter. In the same building, not far from that test stand, Turini and Mariotti had been offered a small desk where they could lodge two computer terminals and follow the data acquisition of their device. Their sharing of test beam resources with Melissa's team occasionally caused some friction: the beam schedule was dictated by the calorimeter group, so the silicon guys had to submit to Melissa's decisions. Furthermore, being confined to that small desk made them feel like second-class citizens. Their attempts to adorn their work post with pictures of pin-up girls from a calendar also caused a small war of nerves with Melissa.

The situation became increasingly frustrating for the silicon team, as the test beam was drawing to a close: it would be terminated in two more weeks, while they had not yet been able to demonstrate the operational status of the readout electronics. Those electric oscillations were killing their hopes. While the work was progressing on many fronts in Berkeley and at Fermilab, Mosé and Nicola concentrated on the circuit designed to damp the oscillations, as they had the feeling that something was wrong there. Finally, one Friday morning, Mosé and Nicola took the box containing the four-layer assembly to the IB4 clean room and opened it. They soon found out that the powering circuit of the SVX chip did not have enough dumping capacity when the charge from the detector required a short burst of current. They fixed the problem by soldering new capacitors in place of the old ones, closed the box, and brought the device back to the Meson area. The chip now worked in sparsified mode with no oscillations!

At the next Monday morning meeting, a beaming Carl Haber got up to show a status report of the silicon detector development and caused several jaws to drop down almost simultaneously in the three conference rooms connected by videoconference from California, Illinois, and Italy. Dante Amidei froze as Carl showed the first meaningful test beam data from the silicon telescope.

"How do you do that???," he asked as he recovered from the shock.

"Well," said Carl, "I simply ran it and it was working."

Dante turned his eyes to Nicola: it was clear that Loose Cannon had fiddled with the detector again.

"So what the hell did you guys do?"

"Nothing much —" grinned Nicola as he replied. "We just stabilized the voltage line with a capacitor. No big deal, but it makes all the difference."

Nicola and Mosé were happy, but now they wanted more: the two remaining weeks of test beam could be used to study the tracking reconstruction capabilities of those silicon sensors. In order to do that, they needed to modify the setup of their test stand. But they had no privilege to access the "controlled access" zone impinged by the beam of hadrons. So they would give false credentials to the control room supervisors. Every time they needed to change something in the detector setup they called the control room, gave the names of colleagues who had access privileges, provided the corresponding ID numbers, and got access to the area.

On one of their after-hours rampages, the two Italians "borrowed" a block of plastic scintillator from one of Melissa's cupboards and drilled a half-inch hole in its center. They then scavenged from somewhere around the lab a copper brick and placed it in front of their silicon telescope, such that it could work as a makeshift beam target. The scintillator was placed between copper and silicon, with the hole aligned with the beam; this way it could work as a "nuclear interaction trigger." The scintillator would detect particles only when pions hitting a nucleus in the copper brick scattered hadrons away from the beam; reading out the silicon in coincidence with that signal allowed to select those nuclear interactions. The goal of Nicola and Mosé was to use the silicon sensors to reconstruct the trajectory of those scattered hadrons. This would show that the detector could correctly determine particle trajectories which, back-propagated to the copper brick, converged to an interaction vertex.

To complete their plan, Nicola and Mosé needed to make sense of the position measurements that they were reading out from the silicon. They asked Riccardo Paoletti, a Pisa colleague who was lightning-fast with code writing, to improvise a tracking reconstruction and an event display on the fly. Riccardo quickly complied. As a result, the next batch of pions from the test beam allowed the collection and reconstruction of events which showed not just signals in the silicon sensors, but real reconstructed tracks seen by all four silicon layers. Those tracks could be fitted together by Riccardo's code to the single point where the nuclear interaction had taken place! This was the Grail they had only been dreaming of in the past weeks: a proof that the readout system worked and that the silicon sensors allowed to reconstruct precise multi-track vertices. In the matter of four days, through ingenuity and improvisation, they had gotten all they could possibly want. But they ended up making a small mistake.

No one had been notified of the modification of their setup in the test beam area — of course, this was their style: Nicola had earned his nickname for a reason. Besides, they were not supposed to be able to do it, since they did not even have the authorizations required to perform a controlled access within the area of the lab traversed by the beam. Unfortunately, on the last day of their nuclear interaction tests, they forgot to remove the copper target after the end of their data taking. Nicola and Mosé had left the system up and running to participate to a party thrown by some colleagues from Pavia University. From the house of the Pavia researchers in Country Ridge they could follow the data by logging on remotely to their data acquisition system through a 14,400 baud modem. Predictably, by the time the party ended, the Pisa researchers forgot to go back to the Meson area to remove the traces of their illegitimate operations. This got them in some trouble, as the calorimeter crew could witness the encroachment when they arrived on site on the following morning. Nicola got quite some heat from Melissa, but that was it — after all, a copper brick did little or nothing at all to the pion beam, as far as the plug calorimeter testing was concerned. Besides, Melissa knew that getting Loose Cannon to follow the rules was a lost cause.

The next day, at the Monday morning meeting, Haber could display the first reconstructed silicon tracks, as clean and straight as they had only been dreamt until then. He proceeded to show event displays with tracks, vertices, and even a graph of the measured *Landau distribution* of charge deposited in single strips of the detectors. A Landau distribution is an asymmetric function that accurately describes the energy loss of a high-momentum particle traversing a thin layer of material. It is the natural expectation for the distribution of the charge collected by the electronics when a charged hadron crosses a thin silicon sensor. The measurement of the Landau distribution had been possible by a rearrangement of the test stand, where, with additional scavenged material, it had been possible to "illuminate" single strips of the detector with the pion beam.

In the matter of a few days, the team had proved to their CDF colleagues that silicon was a winning bet. It was doable, and the precision of its track reconstruction surpassed by one order of magnitude that of the CTC, the wonderful gaseous detector on which CDF relied for charged particle measurements. From then on the project started to sail with tail winds. As for the analysis of the data collected in those days, it was soon published by the scientific journal *Nuclear Instruments and Methods.*

The SVX detector was constructed in 1991 and got installed shortly before the start of Run 1A (1992–1993) in the place of the old vertex tracking chamber, in the very core of the detector around the interaction region. SVX was a cylinder made of four concentric layers of silicon sensors, closely wrapped around the beam pipe in such a way that they covered as large a part of the solid angle around the interaction region as possible. An axonometric projection of the SVX detector is shown in Figure 4.

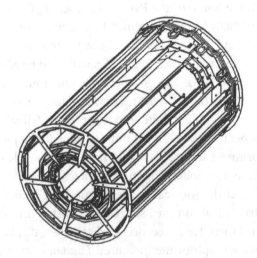

Figure 4. Drawing of the SVX detector, showing one bulkhead and the tiles of silicon crystal ladders assembled in four concentric cylindrical structures.

Summer Student

In the summer of 1992 I arrived at Fermilab as an undergraduate student, full of curiosity about a world I only knew from textbooks. The plan for my three months as a summer student was to take part in a complex analysis which the CDF-Padova group had started less than one year before: the search for top–antitop decays in the all-hadronic final state. I would be working with Esov Velazquez, another summer student from Puerto Rico, under the advisory of Luca Stanco, an INFN researcher from Padova. Luca, a sharp guy in his mid-30s, was spending a full year at Fermilab to get our top search analysis going; he would be our tutor for the Summer Student program. Esov had dark, curly hair, and dark eyes hiding behind light-framed square glasses. He had a strong theoretical background in particle physics, and by hearing him speak you might have been led to think physics was at the very top of his interests. That impression would, however, only last until you saw him roaming around in a white Camaro 25th anniversary, with interiors and front protection lined with mean-looking black leather.

It took Luca a couple of afternoons to give Esov and me a crash course on quantum chromodynamics at hadron colliders, top decay kinematics and selection strategies, Fortran programming, the *VAX* operating system (the language spoken by the computers we would be using), and the dreaded *Analysis_Control*, the framework which allowed one's program to analyze the data. It was an ugly start, but it worked. We got the loud and clear message that we would have to work very hard if we wanted to contribute in some way to the ongoing analysis effort during the three months of our stay at Fermilab.

Besides the specific analysis topics which we would be exploring during our summer student program, one of our satellite assignments was to exploit to the fullest the chances offered by the intellectually stimulating environment. Fermilab was brimming all year long with seminars, colloquia, meetings, and short training courses. Especially during the summer, there was an accumulation of such offers. That was definitely enough to fill up one's agenda, and we needed to carefully select what to attend to and what to skip. The meeting of the Heavy Flavor working group, however, was one we were not expected to miss.

At that time, the working group which organized the searches for the top quark was not yet identified as the "Top group." Until the end of 1992, it remained the "Heavy Flavor" group. The top quark had been considered until recently a "regular" heavy-flavor quark: still much heavier than the heavy charm and bottom quarks, but not qualitatively different from them. That perception was changing due to the high lower limit that the analysis of Run 0 data had set on the top-quark mass. However important, the above considerations did not significantly influence the decision to change the group's name. Instead, the working group was eventually split into two separate groups, the "Bottom group" and the "Top group" by a vote at an Executive Board meeting in November 1992. That was a move necessitated by the huge amount of work that the collaborators were investing in analyses aimed at understanding the phenomenology of each of the two third-generation quarks. The meetings of the Heavy Flavor group had simply become too long, crowded, and complicated to handle. Too heavy, so to speak.

"What's that Chickenshit?"

The Heavy Flavor meeting was held every second Thursday afternoon, during busy "on-weeks," which alternated to "off-weeks" when no physics meetings were scheduled. To me, attending a meeting of the Heavy Flavor group was a real event. I remember very well my first time: Esov and I took the elevator to the right of the main entrance of the Hirise, where we were joined by a dozen people who told one another jokes I did not get (my understanding of spoken English was quite poor back then). When I got off on the 12th floor, on the North corner of the building, we saw the Snake Pit, a square conference room equipped with a large oval table in the middle, and chairs everywhere else. The wall opposite to the entrance was lined with blackboards and a large square screen. In front of it stood a transparency projector, and to its right were placed side by side two large television sets, with a video camera on top. That was the videoconferencing system which allowed remote sites to follow the meeting.

Esov and I arrived a good 10 minutes earlier than the start time. The room was still empty. Upon seeing that around the oval table stood chairs which looked quite the most comfortable we did not hesitate long, and took them. A few minutes later, a tall and lean guy in his mid-30s, better dressed than the average physicist, arrived and took a chair on the opposite side of the table, next to the remote controls of the videoconferencing screens and the slide projector. Turning around on the revolving chairs with curious eyes, leaning back and jolly, we certainly looked more like tourists than physics students. The guy gave us a perplexed look. He opened a logbook, wrote something in it; then he addressed us with the best smile he could manage.

"Good afternoon ... I do not think I have seen you at this meeting before. May I ask you to introduce yourself?"

My English was trembling and broken, but I managed to put together some sort of answer. "Er ... I am Tommaso Dorigo. I am undergraduate student in the University of Padova."

"And my name is Esov Velazquez. I am an undergrad from the University of Puerto Rico. Like Tommaso I am here as a Summer Student, working under the advisory of Dr. Stanco."

"I see. Stanco ... Yes, I know him. Nice guy. What are you working on, then, top physics I guess?"

Esov's English was way better than mine, and I was happy to hear him answer first:

"Yes, we are contributing to a search for all-hadronic decays of the top-quark pairs, in the multijet dataset."

The guy nodded, and he went back to scribbling in his logbook. That would have been the end of it for a less naive person than me, but I felt compelled to continue the conversation:

"... And who are you?"

His perplexed look turned now hopeless, but he kept cool. His tone was, however, slightly stiffer as he explained:

"I'm Tony Liss, professor of Physics at the University of Urbana-Campaign, and I chair this meeting with Claudio Campagnari."

That was my debut at the Heavy Flavor meeting, a demonstration of my utter inability of keeping my mouth shut when advisable. It was only the first of a long series of verbal hemorrhages during my long career in CDF. Thankfully, on that occasion I decided that was enough embarrassment for the afternoon. I made my best effort to act nonchalantly as dozens of other physicists started to swarm in, taking chairs along the walls and leaving the table seats free for more experienced members of the group. The feeling of many eyes staring at me slowly faded away as all the chairs were eventually occupied and the meeting started. Luck had been on my side: as I would later find out, physicists in research labs are by and large not really attached to formal issues; that afternoon, two summer students sitting in the better chairs at the big table did not raise a gram of eyebrows. At the meetings of the CDF experiment, but also during the review process of analyses, the drafting of publications, the exchange of public e-mail messages, and all the other situations when members had to interact, there was a total disregard of academic titles. It was normal for the undergraduate to sit next to the professor. It was just as normal for the former to interrupt the latter during a debate, or at least to show no reverence if a mistake of the professor needed to be exposed. This was a real point of strength of the experiment. There only existed two categories of persons: collaborators, and the rest of the world.

I soon identified the other group convener, Claudio Campagnari. As he introduced the meeting, I stared at him in admiration. In his early 30s, being at the center of the attention in the heart of one of the most important particle physics experiments in the world looked to me like a huge achievement. My opinion regarding the deductions that could be made on the scientific merits of physicists by observing them chair an analysis meeting would change in the forthcoming years. Yet, my respect for Claudio Campagnari, who commanded the scientific output of CDF on heavy flavor physics, would stay high for the years to come.

The introduction was followed by short update reports from the subgroup leaders. In rapid succession, Esov and I could hear the status of searches for a top signal in the dilepton final state, in the single-lepton final state, and then a summary of other analyses of high interest and relevance for the top hunt, which was then in full swing. After those summary reports, a graduate student from Pennsylvania University took the stage to present his work. It was the first detailed talk of the meeting, which reported on the performance of a *kinematic fitter.*

A kinematic fitter is an algorithm which attempts to tailor to a given hypothesis the measured energy and direction of all particles observed in the detector from an event, varying them within a tolerance set by measurement uncertainties. If the hypothesis matches the event characteristics, or in other words if the tailoring succeeds, the fit returns the most likely kinematic configuration of the event, as well as the most probable value of a few derived quantities. The study that the student was presenting used a sample of Monte Carlo events, which simulated the production of top-quark pairs followed by their single-lepton decay. The kinematic fitter was designed to determine the most probable mass of the top quark (known from the start in the used simulated sample, but assumed *incognita*) from the observed lepton and jet measurements.

As the student placed on the slide projector a transparency showing a histogram of the top mass reconstructed by his algorithm, the guy sitting to my left started to complain. It was Ayi Yagil, a tall, lean, neatly dressed guy with dark hair, olive skin, and quickly flying hands which accompanied his speech everywhere it went — very far, sometimes. Handsome, 35 years old, he spoke with a strange accent reminiscent of Hanna Barbera's Yogi the bear.

"Wait a minute, what's that chickenshit scattered around at random masses? Your algorithm is failing big time on those events!"

With his funny expression, Avi was pointing at several bins of the histogram displayed on the screen, each containing very few entries. Those bins were separated from the core of the distribution, whose bins contained tens or even hundreds of entries each. Chickenshit! Those outliers indeed exposed some pathology of the reconstruction algorithm, and Avi was right in his assessment, although he had not formulated the question in a very respectful way. The speaker hesitated, as Avi continued his diatribe using the tactical trick of cutting off the person he was questioning by addressing his speech to someone else.

"You see, Claudio? I am telling you, that is exactly what we should avoid: fitting methods that throw away an inordinate fraction of the signal by attempting a full-fledged reconstruction no matter what, and which cannot be kept under control. Look at the plot: only a small part of the top events cluster at the right mass! The inefficiency is huge! One can't measure anything with it …."

Claudio was a good friend of Avi; together, they had developed a sophisticated algorithm to identify b-quark-originated jets and worked elbow to elbow on the top search. He knew Avi sometimes needed to be restrained.

"Wait, Avi, let him speak," he interrupted.

The graduate student took courage:

"Yeah, I guess the algorithm still requires some tuning …. I will have a look at those events to understand the origin of the fit failures. But I'd like to stress the fact that here I have ran it over the whole sample containing a lepton and at least three jets, without applying the tight cuts we usually make to select the cleanest topologies …."

Avi did not appear placated, and the discussion went on — and on. A similar fate awaited the other talks in the line. By the time the meeting ended, three hours later, Avi had questioned with similar *vis polemica* all of the speakers at least twice. Judging by that performance, my first impression of Avi was that of a guy who seemed to know *everything* better than those who were doing the work. He also looked like one of those naggers with a career who like to use their position and their experience to amuse themselves by shooting at sitting ducks at every opportunity. And yet I had to admit to myself that, in a peculiar way, I found Avi Yagil a fascinating person.

Chapter 5

Run 1

The air in CDF was electric during the hot summer of 1992. The chance to make history by finding the sixth quark inspired and excited old and young researchers alike. The old leaders saw in the top-quark discovery their last chance of catching a Nobel Prize. The younger post-docs and tenure-track professors wanted to show their worth in the challenge, paving the way to better academic positions. Tom Ferbel, a professor at Rochester, recalls a conversation with Paul Tipton, who had just been hired as an assistant professor there. After explaining to Paul that he had overheard his University of Michigan colleague Dante Amidei say he, Dante, would discover the top quark, Tom was staggered by Paul's reply:

"Oh, yeah? Let's see who will be first. I will find top, even if it's not there!"

Tipton does not remember the episode and has doubts he expressed himself in that way. But whether or not the episode played out exactly as Ferbel recalls, somehow it expresses the general feeling: everybody wanted to work on the top search and make a difference. Before they could do that, however, they would have to collect lots of good data.

Testing the Abort System

Run 1 had started on June 5, and physicists were extremely busy tuning the detector components, setting the thresholds of the trigger system, and finalizing the calibration procedures and the reconstruction algorithms. Their only goal was to get plenty of well-understood data to their desktop computers in as little time as possible. It was exciting to see hundreds of skilled individuals working as one, united in the belief that through their hard work they had a chance to produce important advancements for fundamental science.

This time CDF would not be alone in the chase of the top quark. A second collider detector run by a young and combative collaboration, DZERO, had also started operations in 1992. Sitting little more than a mile away along the ring, DZERO was getting the same quality and amount of data that CDF collected, but it was as different from CDF as a collider detector could be. While the design choices of CDF had focused on a precise tracking of charged particles, the emphasis in DZERO had been put on the calorimeter, which was composed of uranium and liquid argon. Liquid argon is a very effective active material for a particle detector, as it produces a good light yield when traversed by ionizing particles. Uranium, in turn, performs excellently three different tasks. First of all, its nuclei are a tight barrier where energetic hadrons quickly convert their kinetic energy into a stream of secondary particles. Second, the strong electric field of uranium atoms is effective for photon conversions and electron bremsstrahlung, which yield well-contained electromagnetic showers. And, last but not least, the unstable uranium nuclei provide extra ionization energy through their radioactive decay when they are hit by particles. By calibrating the response to that extra energy release, one may operate an effective compensation of the collected signal to showers of hadrons, improving the resolution of the jet energy measurement.

The DZERO calorimeter was a technological jewel, but its complex design demands had made it impractical to lodge a solenoid in its interior. Therefore, DZERO was non-magnetic. In the absence of a magnetic field, charged particles produced by the collisions underwent no curvature of their trajectory in the tracking volume. As a consequence, their momenta could not be measured; only their energy could be recorded in the calorimeter. The lack of a chance to cross-calibrate momentum and energy measurements was a clear disadvantage. A second drawback was the lack of a silicon vertex detector, on whose technology only CDF had put its trust. Until the end of Run 1, the DZERO collaborators used to sternly defend the design choices of their apparatus in conference-dinner conversations and cafeteria chats. Yet, later they would show signs of redemption, as the DZERO upgrade for the upcoming Run 2 mainly consisted in the addition of a central solenoid and a state-of-the-art silicon microvertex detector.

Regardless of the above limitations, DZERO was a worthy rival to CDF, and the thought that it could rob the latter of the long-sought-after discovery of the top quark was unbearable to CDF collaborators. Both experiments had to carefully watch over the full operational status of their critical subdetector components, as the loss of data due to hardware problems could provide the opponent with a potentially unsurmountable lead. Despite the redundant measurement of particle momenta granted by the CDF apparatus, a dozen subsystems were critical to the successful online identification of potentially interesting signatures and for the offline event reconstruction. The malfunctioning of even just one component would cripple the resulting datasets, making them unusable for the top-quark search.

One of the critical components for the top search in CDF was the SVX. This was a daring new device, quite unlike the rest of the pieces making up the CDF detector, which used well-tested technology. With its construction, the collaboration had taken a bold step forward, but now it was necessary to keep a close eye on that very delicate system. The SVX was lodged very close to the beam line around the interaction point, so it ran the risk of getting damaged by a shower of stray protons deviating from the Tevatron beam orbit. The electronic chips performing the front-end readout of the silicon sensors, in particular, were not radiation-hard. Beam losses had to be monitored: any time the beam conditions were not considered stable enough to guarantee the safety of the SVX, the latter had to be turned to stand-by. Once de-energized, the system was less exposed to the risk of damage.

Putting the SVX on stand-by under unstable conditions was a manual operation, and as such it might not be sufficiently prompt. In order to provide an automated line of protection from large radiation doses, an active system was installed to monitor the radiation losses from the Tevatron beams. If the radiation level inside CDF went above a configurable threshold, the system was designed to send an abort signal to the

Tevatron controls. The abort signal would immediately dump into suitable absorbing targets the protons and antiprotons circulating in the machine. This was a true novelty: for the first time, the experiment had a way to bail out and spell the ultimate word on the safety of the running conditions, bypassing the experts in the Tevatron control room. After many negotiations, the accelerator division had reluctantly surrendered control to CDF over the fate of their beam. The University of Michigan had built the abort system in collaboration with the Fermilab beams division. It used the continuous measurement offered by beam luminosity monitors mounted on the beam pipe at the two sides of the central detector. The monitors collected ionization charge from stray particles in two-second intervals, providing a real-time measurement of the radiation dose received by the inner part of the detector.

One morning Dante Amidei, who had designed and supervised the construction of the system, was instructing his graduate student, Andy Donn, on the operation of the abort-sending program. They were sitting in front of a terminal in the control room while the Tevatron was starting a new store. By pure chance, the store also promised to be a record one: Run 1A had started just a few weeks before, and the instantaneous luminosity was much higher than the values achieved in the few stores that had preceded it. Dante thought it was a good idea to explain the working of the program while there was beam in the machine, in order to watch the radiation loss readings in real time. In the attempt to explain the working of the dose threshold, he started to modify it.

> "You see, this parameter here controls the threshold of the Tevatron abort system. You move it up and down using these arrows. To show how it works, I'll bring it all the way down to zero."

Andy was not sleeping, and as Dante kept his finger on the down arrow key, he replied:

> "But wouldn't that abort the Tevatron beam?"

At the very same time, the dial of the threshold reached zero. As Dante instinctively looked up to the Tevatron monitor, he saw that the readings of the Tevatron beam currents had zeroed down. Animated voices started to shout across the control room: "We lost the beam!" followed by an outpouring of profanities in three different languages. Dante had chosen a rather awkward moment to test his system, and he had caused it to send the dreaded abort signal. He walked to the shift leader and admitted his mistake. The main control room of the Tevatron was informed of the cause of the abort. Soon the slow accumulation of a new stack of antiprotons was restarted. Besides the loss of a trillion precious antiprotons, the incident did not have other big consequences for the machine.

As it was time for lunch, Dante called it off and got to his car, heading to the cafeteria. When he got out of the car in the parking lot in front of the Hirise, he was spotted and approached by his CDF colleague Hans Jensen, who convened the trigger working group.

"Dan, I saw John Peoples. He is looking for you."

As a punishment, Dante had to agree to the laboratory director's request of giving a talk at the next "all experimenters meeting." This meeting took place every Monday afternoon at 4 pm. It brought together all the experimental physicists working at the various Fermilab experiments, a carryover from times long past when the many ongoing activities in the laboratory were more directly connected to one another. It was agreed that Dante's talk would be titled "How it happened and how it will never happen again." In his speech, he described the accident and the precautionary measures which were immediately thereafter implemented to prevent future screw-ups.

"Is Dante Amidei Listening?"

Soon after the start of Run 1A, a meeting was held to discuss the operational status of the SVX. Many waited eagerly to see how well the silicon performed. One particular concern was the tracking efficiency. The failure to reconstruct even a small fraction of the charged tracks coming from a B hadron vertex could significantly degrade the detector's capability to identify b-quark-originated jets, which was by then recognized as a winning weapon to spot top-quark decays. The detector had purposely been constructed by assembling the four concentric rings of silicon sensors in a way similar to tiles on a roof, such that they would cover as much of the azimuthal angle around the collision point as possible. Like a roof tile, each sensor was slightly tilted sideways in order to overlap with the edge of the one to its right. The masterful technician Mike Hrycyk had recommended that the sensors not touch each other because this would risk damaging the microbonds created between the electronic circuitry on their edges. Some spatial tolerance had to be left between the adjoining sensors. The final layout had been decided in 1990 by Dante Amidei after a long discussion with Aldo Menzione. Aldo wanted to leave as little space as possible, to prevent charged particles from escaping undetected through the space left between the sensors. He actually insisted on a design which foresaw gaps as narrow as one micrometer! Overruling Aldo's requests, Dante had chosen a way more conservative arrangement.

At the meeting, followed via videoconference from Italy, Michigan, and Berkeley, a researcher showed the first graph of the measured tracking efficiency of the SVX as a function of azimuth. This was computed as the fraction of well-reconstructed trajectories measured in the CTC which failed to match to one of the track segments reconstructed in the SVX. Overall, the graph had the properties that everybody was hoping for: a rather high tracking efficiency for all values of the azimuthal angle. This meant that all the components of the detector were functioning normally. However, some narrow regions showing 5% deficits were evident at regular angular intervals. The regions spatially coincided with the sensor boundaries, where particles appeared to escape undetected. The speaker was just starting to discuss his results when a loud baritone voice with a strong Italian accent resounded in the room from the videoconferencing system. Each word was precisely articulated so as to surmount the static background of the system and the evident language barrier that existed between the owner of the voice and the person he was addressing.

"Is--Dante--Amidei--listening?"

That was Aldo, who was following the meeting from Pisa. He had not forgotten his argument with Amidei about the placement of the silicon sensors, and he wanted to finally make his point with the support of hard data: the space left between the tiles caused a loss of good tracks! However, it later transpired that the loss had another cause. The inefficiency at the edge of the sensors was rather due to the use of a still imperfect algorithm for the reconstruction of the clusters of ionization charge in the strips. Each layer of the SVX was instrumented with sensors that had narrow strips parallel to the beams. When a charged particle emitted from the interaction point crossed the system, it left ionization charge in each layer. The charge was collected in clusters of a few adjoining strips. An algorithm called "cluster finder" found the barycenter of the charge deposition on each layer. That information was then used to reconstruct a track segment by connecting the four resulting spatial points along the particle path. Now, when a particle crossed the region where two sensors overlapped (the spots at the heart of the dispute between Aldo and Dante), it could deposit charge on both sensors, which laid at different distances from the beam axis; the special geometry at the edges called for a special treatment in the determination of the barycenter of charge deposition, which the original cluster finder had not been designed to perform. That was the real source of the small inefficiencies.

Muslim Riots

Besides the correct operation of the various hardware components, a successful data collection required the correct and continuous operation of many different online programs. Some were mere monitoring routines which verified the operation of detector elements and performed simple tasks. They could report warnings to the control room crew when something smelled fishy, or take more extreme measures like aborting the Tevatron beam as mentioned above. Other programs attended to more complex duties, such as the continuous checking of the data as they got collected, or the extraction of calibration constants required to perform corrections to the online detector readings. Those corrections ensured that the trigger would operate decisions based on accurate energy and momentum estimates of the identified particles and jets. Without them, a change in temperature or currents, a slight modification in the magnetic field of the solenoid, or other transient effects could worsen the performance of the whole data acquisition.

One critical program was called *Stage0*. Stage0 had been written by Aseet Mukherjee, and as was typical of Aseet's work it was a quite complex piece of code. It operated iterative corrections on the data read out from the central tracking chamber (CTC), based on calibration constants derived from the previous run. This had the purpose of accounting for the potential drift of the magnetic field and other issues affecting the CTC readout in the course of data taking. That information was used by Level-3 triggers which collected events with high-momentum particles. Although the datasets which depended the most on correct Stage0 calibrations were not directly used in the top-quark search, some of them allowed for important background studies of relevance for top-quark analyses and for the experiment in general.

As he used to do during the fall every other year, in October 1992 Aseet left with his family for Kolkata, the Indian city where his parents lived. Unfortunately, a week before Aseet's planned return, Stage0 started to send ominous-looking error messages to the crew in the control room. A quick investigation showed that a previously undiscovered bug in the code was causing a corruption of the files where Stage0 wrote the CTC calibration constants at the end of each run. The program was unable to read back those values at the start of the new run, so it could not perform its task effectively any more. As a result, the Level-3 track triggers acquired data of increasingly poor quality as data taking progressed.

The CTC experts Peter Berge and Robert Wagner were summoned to examine the complicated code of Stage0. They could not find any easy-to-fix bug. They also decided it would take too long to rewrite the program from scratch: in a week Aseet was due back, and he would be able to solve the problem much more quickly and effectively. Yet, the situation was worrisome: the Level-3 track triggers were now collecting data that would prove unusable for offline data analysis. Meanwhile, Aseet could not be reached by phone or other direct means, which made everybody even more nervous.

While Aseet was packing his luggage to head back to the USA, violent riots broke out in Mumbai and Kolkata. The friction between muslims and hindus is a long-standing problem in the region, and bloodsheds are recurrent there. Soon the situation in Kolkata became simply unsafe for travel, and flights were canceled: Aseet was stuck in India! Peter Berge managed to partially patch up the code of Stage0, but even his magic powers were not sufficient to put the routine back in perfect working order; the resulting data were not good enough. The whole collaboration waited for Aseet's return in a state of increasing anxiety.

Finally, after two weeks of delay, Aseet managed to fly to Berkeley; from his home institution he could connect to the CDF computers. He soon managed to correct the bug. He then initiated the long process of spinning all the datasets collected in the course of the previous three weeks, iteratively recomputing all calibration constants and restoring the functionality of the Stage0 corrections. But the accumulated stress took its toll: one evening he collapsed in his office. An ambulance brought him to the ER, where he was diagnosed with pneumonia. As he received the first medications, he informed his CDF colleagues of his sickness. The further delay was not welcome by the managers, who put pressure on Aseet to return to Fermilab immediately. It took Lina Galtieri, the head of the Berkeley group in CDF, to support Aseet in his decision to wait and give priority to his health over the needs of the experiment.

Three days later, a still ailing Aseet eventually flew to Chicago. The calibration files were soon fixed, and CDF went back to its full data collecting potential. Following this incident, on the insistence of the head of the Particle Physics group, Aseet would take a portable phone with him in his future trips to India. He was also explicitly forbidden to fly in the company of Robert Wagner, the other tracking guru of CDF. To paraphrase Oscar Wilde, "losing one of the two CTC experts would be a tragedy, but losing both would be reckless." Never again would the CDF knowledge base of the tracking software be thus jeopardized!

The Trigger, a Battleground

As was mentioned in Chapter 2, during Run 1 the trigger system selected at most 50 events every second, which were then stored in a magnetic tape. It did so by discarding in real time 99.98% of the collisions produced in the core of CDF, by applying a set of requirements on measured quantities of interest. Writing to tape the full detector information 50 times a second was roughly the limit of the data acquisition. The real puzzle was how to make the best use of that capability, defining selection criteria capable of sorting out the most interesting collisions and discarding all others. The highly complex menu of selection criteria applied by the trigger included literally hundreds of different ways (called *trigger paths*) through which events could make it to data storage. The different trigger paths responded to the needs of different analyses and searches pursued by CDF, each path receiving a different share of the total output rate: a different *trigger band-width*. Events were collected if they contained an energetic electron, or an energetic muon, or large missing transverse energy, or four energetic jets. Each of the above was a possible signature of top-pair decay, and CDF had to collect all the events of that kind. Yet, many other combinations of objects identified by the online reconstruction system were needed by exotics searches, electroweak measurements, studies of QCD, or B hadron physics. Compromises had to be worked out on the bandwidth to be allocated to each of the trigger paths selecting those events.

The definition of the selection criteria of the various trigger paths was not a problem which once solved would stay solved. The Fermilab scientists and technicians who operated the accelerator were doing their best to continuously increase the number of protons and antiprotons circulating in the Tevatron ring. That made each of the 280,000 crossings of proton and antiproton bunches in the core of CDF more and more likely to produce interesting collisions. A higher luminosity meant a higher sensitivity to rare processes and a better discovery reach. However, because of the increasing probability that bunch crossings produced collisions passing the trigger requirements, the trigger accept rate was bound to increase. Simply put, a trigger selection which produced an average output of 50 events per second at a given beam luminosity would yield 100 events per second if the luminosity doubled. The trigger settings would then have to be modified. Otherwise, like a funnel receiving too large an influx, the data acquisition system would overflow, washing out all benefits of the increase in luminosity. Every second event would be lost, as if the system were dead 50% of the time. With the ever increasing Tevatron luminosity, the high-rate triggers collecting hadronic jet events were usually designated as the sacrificial lamb: they were not critical for the top search.

The Trigger group meeting was the crowded arena where executive decisions were routinely taken on what data to take, how to set the trigger thresholds, and how to watch and reduce the system's dead time. It was held every Wednesday afternoon in the Pump Room, a small conference room adjacent to the control room on the second floor of the B0 building. There I gave my debut as a speaker in CDF on December 2, 1992. The agenda on that occasion had been built around the issue of how to reduce the dead time in the data acquisition system produced by the recent increases in beam luminosity. That afternoon, I would learn that CDF was not always a place where controversies were resolved by the slow and painful formation of a consensual view. Sometimes, urgent decisions were required, and this involved the exercise of authority.

With the help of my thesis advisor Luca Stanco, I had studied how the collection of fully hadronic decays of top-quark pairs (ones only containing hadronic jets; see Chapter 4) would be affected by a proposal put forth to reduce the rate of accepted events and thus hopefully cut down the dead time to an acceptable level. The proposal involved bumping up to their maximum value (51 GeV) the transverse energy thresholds on localized readings in the plug and forward calorimeters. Those thresholds were used by the Level-1 trigger to accept events with forward jets, as the latter left enough energy in those localized regions to pass the trigger selection. By studying the characteristics of simulated top-quark decays, I had come to the conclusion that only a very small fraction of reconstructable hadronic top events relied on the presence of forward jets to pass the Level-1 selection. In other words, top-pair production events were special enough that they would get collected anyway by other triggers; the implementation of the proposed change would not negatively affect our analysis.

My talk was the last one of several reports that had focused on the issue from different viewing angles. After I finished speaking, a heated discussion ensued around the big table at the center of the room. Melissa Franklin argued against the change. As mentioned in Chapter 4, in the past few years she had directed the commissioning of the plug calorimeter, which measured forward jets, and she was now using the early Run 1 data to study the performance of that system. The proposed change in threshold would seriously hamper her analysis. Others were also unhappy at the prospect of losing forward-jets data. One was Fermilab associate scientist Brenna Flaugher, who was at the time about to become a convener of the QCD group and was in the process of measuring the angular distribution of jet pairs. A few additional collaborators opposed the proposal by arguing that the higher thresholds would not even be sufficient to reduce the accept rate to safe values, and suggested that further studies had to be undertaken. Others suggested sacrificing other expendable data samples.

The Trigger group leaders looked unmoved by the objections. They considered that the main priority was to preserve the experiment's chances to collect as many top-quark events as materially possible. They had verified that their opinion was shared by a large majority of the collaboration. A decision was urgent, since until one were taken, the CDF data acquisition system would continue to suffer a dead time of up to 30%. The dead time constituted a horizontal cut of 30% on all datasets across the board. Three out of 10 collectable top-quark events were going down the drain!

Rather than moving toward a compromise, the discussion became contentious, with physicists growing more and more radical in the expression of their own opinions. And this was in spite of the strict scientific rationale of the issues presented, on which they would have surely had to agree. The depth of disagreements had to do with politics, specific interests of different research groups, and personal rivalries. From behind a thick brown-and-white beard, Hans Jensen, the strongly built veteran of collider physics who chaired the meeting, finally cut it short.

> "I think we need to close this discussion here. I have heard all of your objections but it seems to me that there is really no alternative to raising the plug and forward thresholds: any decision is better than no decision."

Indeed, taking no decision would keep the dead time unaltered, hampering the discovery potential and the measurement precision of every single analysis. That would be a democratic, but also a rather meaningless and unscientific way to manage the data taking of the experiment.

Melissa did not find Jensen's words convincing. She thought that the management was being stubborn and that there were better ways to handle the emergency. She rose from her chair and continued to argue as she walked around the large table in the direction of the door, explaining once more that there were alternatives to killing the chance of calibrating the forward calorimeters. Somebody in the audience counter-argued that CDF had not been built with the primary goal of studying the physics of forward jets. As the volume of voices suddenly rose, Melissa understood that the battle was lost. So she wished aloud a four-letter word to everybody and left the room, imparting to the door slightly more kinetic energy than that needed to close it. Melissa's scorn was understandable: her reasons to disagree with the decision being taken were sound, but they were not being valued as such by the management. The top-quark search was the absolute priority of the experiment. This was specifically clear in the mind of the managers, who organized the effort of all collaborators to maximize the chances of a discovery. Fortunately, similar arguments, even if bitter and heartfelt, did not significantly affect the general climate of cooperation within the experiment.

B Tagging

In addition to babysitting the detector and the data acquisition system, at the start of Run 1 physicists spent their time refining the tools with which they would reconstruct and identify the salient features of top-quark decays. Separating the signal from the very large backgrounds was a critical aspect of the search. In fact, the Run 0 top search had shown that the top production rate was very small, so a top-quark discovery would have to rely more on the dirty but relatively more frequent single-lepton decay rather than on the clean but rare dilepton decay.

As discussed in Chapter 4, when the top quark decays it usually turns into a bottom quark by emitting the charged-current weak interaction carrier, the W boson. Bottom quarks, although not nearly as heavy as top quarks, have a sizable mass — more than 4 GeV. Because of that, they are much less frequently produced by QCD processes than lighter quarks, the featherweight up, down, strange, and the medium-weight charm. If one takes a random sample of generic hadronic jets produced in Tevatron collisions, one may expect that on average only about half a percent of them have been originated by b-quark hadronization. Hence top decays can be successfully distinguished from generic QCD processes by the presence of b-quarks in the final state. By virtue of this difference, bottom-quark-jet tagging methods were soon recognized to be the most powerful way to discriminate top-pair production events from all processes capable of mimicking the same signature of leptons and jets.

Back in Run 0, the only available *b-tagging* algorithm was one which sought for low-energy electrons and muons embedded within hadronic jets. Bottom quarks decay to those particles about 20% of the times, while lighter quarks and gluons almost never feature similar characteristics. This method of identifying bottom-quark-originated jets was called *soft-lepton tagger*, or SLT. SLT was a software algorithm written by Claudio Campagnari and Avi Yagil; it combined the information from all the subdetectors which could discriminate electrons and muons from pions and other hadrons. Jets containing good lepton candidates were tagged as b-jets; jets without leptons were classified as "untagged." Requiring the presence of one SLT b-tagged jet in W+jets events reduced backgrounds by over an order of magnitude, while retaining about a fifth of the original top events in the sample. This enhanced the sensitivity of the search.

The electron-finding code in SLT was Avi Yagil's creation. The optimization of the algorithm took a lot of time and effort. For a while Claudio, who was handling the muon identification part of SLT, kept bugging Avi to wrap his part of the job up, but Avi wanted to do more calibration studies. So Claudio would walk to Avi's office, yelling at him: "Are you done yet??!!!" Huge verbal fights would start, with loud voices that could be heard from one end to the other of the long trailer corridors. To the amazement of bystanders, after the end of the name-calling the two contenders peacefully walked away together to have a coffee at the third floor of B0, or a beer at the Users' Center if the afternoon was over.

When Avi was finally done, he showed the detail of the electron identification in hadronic jets at a Heavy Flavor meeting. The nature of electrons could be inferred by the signal they left in the electromagnetic calorimeter. SLT enforced tight matching requirements between the trajectory of the charged track produced by the electron candidate and the position of its energy deposit. The latter was precisely determined by the CES detector, a strip chamber embedded in the electromagnetic calorimeter at the depth where the electron shower had its maximum transverse extension. For that reason, the detector was also called "shower max." As Avi showed a graph detailing the level of compatibility of track and cluster positions, he was promptly interrupted by Barry Wicklund, who had literally built parts of the CES detector with his own hands.

"Avi, do I read it correctly that you are requiring the extrapolated track to miss the cluster by at most 0.1 centimeters?"

"Sure, it's a cut loose enough that we don't lose efficiency on it. The positions are very precisely measured," said Avi.

Wicklund insisted:

"But that can't be true. I measured the strip pitch of the shower max myself with a ruler once, and the true number is not what is in the geometry database — it's off by at least one millimeter. We never bothered to put in the database the correct number, because we thought that kind of precision was not necessary!"

Avi was unimpressed.

"So what? It is calibrated, and my conversion sample shows that it is working fine."

Avi was referring to a validation study he had carefully carried out on a large data sample of pure electrons. The way to get a pure sample of low-energy electrons was well known to Barry, who had been the first one to use it (see Chapter 3): it was the study of the so-called "conversions." When a photon interacts with atoms of gas or wires in the tracking chamber, it sometimes "converts" into an electron–positron pair. The two particles can then be tracked all the way to the calorimeter and be measured there; the null invariant mass of the pair guarantees that the parent of the pair is indeed a photon. (In principle, photons of sufficient energy may convert into any charged particle pair, but the chance that they do so decreases with the particle's mass, so for all practical purposes one may consider conversions to yield just electrons.) If one takes the pains to select electrons from identified photon conversions, one may then fine-tune the cuts applied to select those particles. Avi had done that study and he knew what he was talking about, but Barry was not convinced.

"Look, Avi, the way the code gets the strip position is to multiply the strip number by the strip pitch in the database. So if the pitch is off by a millimeter, you are increasingly off in your matching cut. By the time you get to the twentieth strip your cut completely misses the true electrons!"

Avi coolly explained:

"Of course Barry. I know about this, because I plotted the residual difference between extrapolated track position and shower max position in the conversions sample. It does show an increasing trend with strip number. So I corrected for the effect of your incorrect database numbers and now it works!"

This took Barry and the rest of the audience aback. From that point on, nobody doubted that the lepton tagging algorithm had been tuned with care.

SLT remained a very useful algorithm for top-quark searches even after Run 0, but starting in Run 1 CDF was endowed with the SVX, a detector which enabled more powerful b-tagging methods. The ionization deposits in the SVX strips left by charged particles allowed the software to reconstruct trajectories with micrometric precision in the vicinity of the collision point. This could be used to detect the decays of long-lived B hadrons within hadronic jets, as the decays of B hadrons produce charged particles from a point usually displaced by several millimeters from the interaction point. The back-propagated tracks of those particles do not intersect the primary vertex. The minimum distance between primary vertex and track, called the *impact parameter*, is the relevant information to distinguish particles originated at the interaction vertex from ones produced by a decay in flight (see Figure 1). The problem was how to make the best possible use of that precisely measured quantity.

Years earlier, when the SVX was only a blueprint, the issue of how to best use its high-precision tracking capabilities had been an important one.

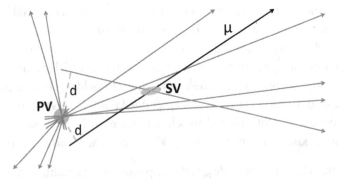

Figure 1.   Bottom-quark-originated jets (one shown here by tracks pointing toward the right) can be tagged by studying the trajectories of the particles they contain. The collision point, or "primary vertex" (PV) is found as the most probable point of origin of all trajectories (shown by arrows). A "secondary vertex" (SV) spatially distinct from the PV can then be sought using tracks within a jet that have non-zero impact parameters (labeled by the letter "d" and shown by dashed lines). The identification of an electron or muon (arrow labeled "μ") in the jet provides an alternative b-tagging method, SLT.

Run 0 data showed that CDF had a high potential in the nascent field of B physics measurements and it started to become clear that the SVX would further advance that potential. Yet, Dante Amidei and collaborators had continued searching for a center-piece application to new particle searches. Soon Dante came across the slides of a 1986 talk by Aldo Menzione, where Aldo had suggested using SVX information to improve the discrimination of top-quark events from backgrounds. As mentioned in Chapter 4, Aldo's idea had been to veto events where electrons or muons displayed a significant impact parameter. Dante understood that he could invert that strategy, now that the top quark was proved to be heavier than W bosons. Heavy top pairs produce b-jets, so using impact parameter information one could *select*, rather than discard, events with B hadrons. As he bounced the idea off Aldo, Dante initially had a very hard time. Aldo was confused: his own plan of vetoing b-quark jets was so ingrained it took him a while to understand Dante's point. Later Amidei showed at the Heavy Flavor group meeting a study of how a secondary vertex could be found in b-jets from top decays, with some 20% efficiency. The presence of a well-reconstructed secondary vertex in a jet is a powerful way to distinguish b-quark jets from jets originated by light quarks or gluons, which background events are rich of. A way to improve the selection of top events had thus been found. Many collaborators invested their efforts in devising the most effective method to distinguish b-quark jets from ordinary ones using the precise SVX tracking information.

In the end, two SVX-based b-tagging algorithms stood above all others thanks to their better performance in terms of background rejection and signal efficiency. The first one, called JETVTX, selected among the tracks in a jet those which possessed a significant impact parameter with respect to the point where the original proton–antiproton collision had occurred. The size of the impact parameter determines the validity of the hypothesis that the particle comes from the primary interaction, just as the distance between the bull's eye in a target and the point where the arrow hits is a measure of how good the bowshot is. Unlike the arrow's trajectory, however, a particle track is measured only indirectly, from the hits it leaves in the four layers of the SVX and in the sensitive wires of the CTC outside it. The measured track, therefore, carries some uncertainty, which depends on the number and quality of the hits it produced, its curvature, and other characteristics.

The trick was to select tracks whose measured impact parameter was much larger than its own uncertainty, indicating it was incompatible with having a null value. JETVTX then fitted those tracks together to find their hypothetical common origin. This "secondary vertex" was a good determination of the space point where a B hadron had disintegrated.

The other algorithm was called JETPROB. Rather than explicitly searching for a well-displaced secondary vertex, which could be problematic to identify when too few charged particles were emitted from the decayed B hadron, JETPROB computed a global probability that the set of tracks belonging to a jet all came from the primary interaction vertex, using as input the measured track impact parameters and their respective uncertainties. A small probability was an indication that the single-vertex hypothesis was wrong and that the jet was likely to contain the decay products of a long-lifetime B hadron.

In summary, starting in Run 1 CDF had two new powerful weapons with which to look for top-quark decays: a b-quark tagging algorithm based on the direct reconstruction of its decay point, and one based on a more efficient, although less pure, selection of jets "smelling of lifetime." Both could identify bottom-quark-originated jets regardless of whether the quarks decayed to leptons, which was a taxing prerequisite of the SLT algorithm.

Software Wizardry

One of the developers of the JETVTX algorithm was Gordon Watts, then a graduate student at Rochester University. Gordon was a lean guy who carried straight blond hair down to his shoulders. Working on the b-tagger day and night, he had grown skilled at using intensively the scarce computing resources of the Heavy Flavor working group. To determine the algorithm performance he needed to study Monte Carlo simulations of top-quark events, as well as large samples of real collision data. Using the rate of b-tags found in the two datasets, he could work out an estimate of the enhancement in signal fraction that the selection of b-tagged jets provided.

Handling large amounts of data was problematic: one needed high-capacity hard disks and lots of computing power. In order to maximize the exploitation of the available computing power, the Offline group networked the workstations and storage disks of the users participating to each working group. Group members could then share datasets and programs with other users, write data to their disks, and run jobs on the workstation sitting on somebody else's desk when its CPU was free. It was a sort of virtual communism; all users had equal access to the group resources. Yet, there were users who, thanks to their higher computing skills, managed to game the system and exploit it better than others. Gordon was one of them; he used to submit to the batch system a script which parallelized the heavy task of running on a large dataset, breaking it into smaller jobs. Each job spun some fraction of the data and deposited the temporary cumbersome output files on the "scratch area" of an available disk — the part of the disk specifically dedicated to store large temporary files. At the end of the process, the relatively small output resulting from a further analysis of the temporary files could be collected and deposited in its final destination.

Gordon's scripts once got the system administrator Richard Krull angry at him, as the large temporary files left behind clogged many disks' scratch directories. Since then, Gordon included in his scripts certain commands which ran at the end of the job, searching and deleting all the temporary files from the disks where the files had been temporarily located.

One morning in the fall of 1992, Flavia Donno arrived to her office in the CDF-Frascati trailer and started reading her e-mails. She was suddenly interrupted by Paolo Giromini, pale in his face, who cried for help.

"Flavia, there's something bad happening to our data files. They are disappearing from our disks!"
"What do you mean?" inquired Flavia.
"I mean I was running a job and it crashed, and I found out it did because the data files it was trying to access aren't there anymore. And since I first looked, I saw that more files are disappearing!" explained Paolo.
"Let me see. There are several jobs running on the cluster at the moment.... I will need to see what each of them is doing."
"Please be quick, this is driving me mad!" begged Paolo.

Those were days when hard disks were not backed up by automatic systems: one had to manually back up one's files to 8 mm tape cartridges to be safe, but the procedure was long and cheerless. Many researchers preferred to optimistically rely on the stability of their hardware and on the absence of exogenous catastrophes. As Giromini's disks were not backed up, the matter was urgent. Flavia started to feverishly track down the action of every running job, using her access privileges. Soon she found a program which was deleting whole directories from one of the disks. The lines of code startled her. Rather than going into a specific directory of a specific disk to delete the files stored there, as one would normally do in order to clean up temporary data stored at a particular location, the script deleted whole directories recursively, without checking at what level in the tree they were located. This was possible as the operating system allowed users to define "logical names" for physical directories. A command using a logical name would then work seamlessly anywhere in the cluster if all disks contained the same directory structure. Flavia immediately killed the rogue process, but soon Giromini, who was checking his archives from another workstation, shouted:

"Here! Here! There're other files disappearing from this other disk!"

Flavia needed to find and kill the parent script which was spawning the directory-deleting jobs, so she checked out the name of the submitter of the rogue processes. The username of the submitter was Watts. As soon as Paolo learned the name of the offender, he started to entertain homicidal thoughts. While Flavia proceeded to kill all the running jobs, Paolo stormed to Watts' office, but found it empty, so he dashed to the office of Watts' advisor, Paul Tipton.

That morning, Gordon arrived a bit late to work, as on the preceding evening he had been working until late. He parked his car on the west side of B0, picked up a coffee at the third floor of B0, and proceeded to the CDF trailers on the other side of the building. As he entered the corridor of the trailers, headed to his office, he bumped into Paul Tipton, who was going to B0. But he got no cheers from him, only a dry sentence, spelt with Paul's usual calm voice:

"Paolo Giromini is looking for you. And you're on your own on this one."

Gordon was perplexed. What did his advisor mean? Scratching his head, he proceeded to the CDF-Frascati trailer, where he met his Italian colleague Andrea Parri in the corridor. Andrea was one of Paolo's collaborators.

"Hello Andrea, how is it going?"

Andrea's smile was a bit wider and slightly more ominous-looking than normal as he answered joyfully:

"Oh, hi there. I'm doing great, thanks! Ah, Paolo is looking for you ...."
Gordon replied: "Yeah, I'm here for that, is he in?"

Paolo heard the voices and got out of his office at the end of the corridor; he reached them in no time. He was red in his face, from his hair all the way down to his neck. Meanwhile, Andrea opened the door of the office he shared with Flavia, whispering to her "It's going on now, come out, don't miss it!" Gordon instead had no clue yet of what was the matter:

"Hello Paolo, were you looking for ..."
"Give me 600 dollars."

Giromini's voice was apparently calm, but his body language betrayed that he would rather use his hands than his mouth.

"What?!?"
"Last night your script deleted my data files and all the macros and programs I wrote in the last few years of work ...."

The student was flabbergasted.

"You can't be serious!" he objected.

Giromini continued:

"... so I have nothing else to do here at the moment, and am going back to Italy. But my ticket's return date can't be changed; I need to buy another one. You owe me the money to buy it. So give me 600 dollars."

Gordon Watts remained stuck, his mouth open, and his face a palette of shifting colors. It was dawning on him. Could it be that the command with which the script cleaned up the mess of temporary files in the scratch directories was being issued one directory or two too high up in the tree? Yep, that was the simple, terrifying explanation.

Gordon felt like crap for a while because of the incident. He did not pay the flight cost for Giromini, but his acrobatic data-spinning scripts were forced to an early retirement. Arguably, it was not entirely his fault. The bug in his script had been there for a while, but had caused no harm until then as most disk areas in the cluster had protections to prevent similar mishaps; the deletion command simply got ignored there. The Frascati disks, however, contained unprotected data and programs in their scratch areas, as this made it easier for the group members to exchange common resources; the researchers had not created a "group area" as other groups had instead been careful to do. This was an unsafe practice, but physicists were accustomed to working their way around the scarcity of available computing resources and the iron rules of the system administrator. And CPU power was as much in demand as disk space. Those who had access to large arrays of processors, or had more familiarity with the batch system, would one day be the first ones to see top candidates appear in well-constructed histograms or scatterplots on their computer screens.

And indeed, by the end of 1992, events smelling of top-quark production started to appear. The process through which those events were gradually recognized to be the true signal of a new quark is a textbook example of how hard it is to build a solid consensus within a large group of people rich with skeptical and free-thinking minds.

Chapter 6

Top-Quark Battles

As Run 1 data started to fill the Exabyte storage tapes in the vault of the Feynman center, the CDF collaboration displayed the first symptoms of a sort of superiority complex with respect to the rest of the world of high-energy physics. Surely, the LEP accelerator was a giant marvel, and the exquisite measurements of the properties of the Z boson that the four CERN experiments had been producing in e^+e^- collisions since 1989 were an impressive achievement. Yet, the Tevatron was the highest energy collider: here laid the forefront of research in particle physics. And it was CDF who would tell the world whether there was a sixth quark; CDF would discover whatever new physics was waiting at the high-energy frontier.

… Or was it going to be DZERO? Alas, CDF was no longer alone at the top. DZERO was a younger experiment: the detector was younger and so were the scientists who manned it. Many in CDF assessed the competitor as clearly inferior in design and performance. What remained to be done was to prove it with hard data, and this was a source of pressure on the senior scientists who constituted the old guard of the collaboration. There could be no false step; no scientific claim could run the risk of having to be retracted. The competition with DZERO had to be won with a scientific output of better quality. As a result of that mindset, the CDF collaboration started to act in a conservative way. This conservativism manifested itself in two ways. The first was the setting of exceptionally high standards for all publications: all physics results had to be of the highest quality, even if that meant an overly tedious and lengthy internal review process.

The second was a tendency to discourage data analysis strategies considered "too aggressive," or simply too original.

It was along the path to the top discovery that the collaboration also experienced for the first time the centrifugal forces that the above-mentioned attitude generated. Until the start of Run 1, scientists from different countries had by and large worked in complete harmony, as they had the common goal of building and making operational the best detector they could put together. Now that the time had come to reap the benefits of that long building effort, the divisive potential of the high prize finally at reach started to play a role. Part of it was the natural desire of power and prestige, to which nobody was completely immune. The rest were the different intimate convictions of physicists with different professional background and experience. Reasonable individuals may disagree, and CDF physicists did so frequently, especially when the issue was as serious as the top discovery.

Luckily, neither in the occasion of the top search nor in several other cases that followed it did the collaboration break under the action of those centrifugal forces. The melange of different cultures and education and the huge intellectual richness of the set of 500 skilled scientists kept the experiment united. The search for consensus always prevailed over personal views, although in some cases this proved very difficult. In this chapter, I will describe some of the clashes and internal battles that had to be overcome in the path to the top discovery.

Jackpot: The Single-Lepton Final State

Among the three main final states of top-quark pair decay, the dilepton one raised no controversies: all that was needed was a simple selection requiring the presence of two high-energy electrons or muons and significant missing transverse energy, plus two energetic jets. It was not even necessary to further increase the sample purity by requiring that the jets contain b-quark tags, as the above characteristics already yielded an expected signal-to-noise ratio as high as four-to-one. The problem was the very small rate of candidate signal events, which made the dilepton search a game of patience. On the opposite end of the purity scale stood the all-hadronic final state. There was no dearth of events in that case: even a very stringent selection including JETVTX b-tagging would collect thousands of events with six energetic jets. The large majority of those events came from QCD processes involving light quarks and gluons, a background very hard to reject. A credible method to spot the top-pair signal in the all-hadronic final state had not been devised yet, and few in CDF believed that it was at all possible.

The golden channel for the most ambitious top seekers was the single-lepton final state, where the signal rate was not too small and backgrounds were not too high. Dozens of physicists devoted all their research time to figure out the best way to identify a significant amount of single-lepton top events. There, the conservative "simple is best" creed was the mainstream and had the support of the leaders of the Heavy Flavor group. A simple selection of the so-called "W+jets" events, featuring a leptonically decaying W boson accompanied by three or four hadronic jets, one or two of them b-tagged by JETVTX, appeared the easiest way to extract a single-lepton top signal. The most challenging part of the analysis consisted in estimating as precisely as possible the background rate: an excess of events due to top-pair production would be easier to demonstrate if the background expectation had a smaller uncertainty.

One method to size up the background relied on measuring the number of b-tagged jets in data purposely selected in such a way that they were devoid of a top contamination. The rate of b-tagged jets in this "control sample" could then be extrapolated to the W+jets sample. By assuming that jets got b-tagged with equal frequency in the control and signal samples, one could estimate the number of b-tagged background events in the latter from their rate in the former. This method was imaginatively called *Method one*. A second method consisted in carefully tuning the Monte Carlo simulation of all known reactions producing the W+jets signature, and then relying on the Monte Carlo prediction for the number of events that each process yielded in the final data selection; to distinguish it from Method one, it had been imaginatively decided to call it *Method two*.

Method one was data-driven and was therefore regarded as the safest. Monte Carlo programs were in fact not trusted to give a correct prediction for the rate of the complex reactions yielding many hadronic jets. Indeed, that kind of reliance on simulations had been a mistake of the UA1 experiment less than a decade before. That incident was still vivid in the memory of many CDF members. In 1984, the UA1 physicists had trusted too much the simulation of W production. One year after the giant success of the W and Z discovery, the data were mined in the hope of ticking off the top discovery as well. A handful of events were observed which featured energetic leptons and additional hadrons consistent with the decay of a high-mass object: a possible signature of a 40-ish GeV top quark. Simulations did not predict that the W boson would be accompanied by so much hadronic activity. The UA1 collaborators came to the conclusion that those events were due to top-quark production and published their analysis. The claim was later retracted.

In the light of the UA1 experience, CDF was determined to be extremely cautious. The UA1 analysis was otherwise sound, but had been betrayed by an incorrect Monte Carlo prediction, on which it made too much reliance. Hence, the CDF Method-two estimation of backgrounds with simulated samples was only accepted as a cross-check of the background prediction computed with Method one. Yet, it was an important check; it could be argued that the technology of computer simulations had made great progress in the past decade. Now physicists could rely on a tool designed specifically for the simulation of the dominant W+jets backgrounds: *VECBOS*. VECBOS had been written by Fermilab physicist Walter Giele in 1989. It was a Fortran program that simulated all the possible hard processes leading to the creation of a W boson accompanied by energetic quarks and gluons in the final state. Each event was produced along with a weight: a number which kept track of the relative importance of the particular kinematic configuration that the event represented. If an event had a weight five times larger than another, it meant that the former occurred five times more frequently in real collisions. The weight accounted for the non-uniform way by means of which VECBOS scanned the parameter space of the production processes.

Generating large event samples with VECBOS was a real pain. The computing power required to effectively scan the grid of points thrown by the simulation program to sample the parameter space was enormous for those days. Furthermore, the output of the program depended in some measure on the choice of initial settings that the user provided. Those arbitrary choices were a source of systematic uncertainty on the result.

It was thus necessary to simulate samples with different initial settings: a comparison of the outputs would tell how significantly the specific choices of the settings affected the final result.

In 1991, Claudio Campagnari had succeeded to Brig Williams as co-convener of the Heavy Flavor group. He was spending 110% of his time on the top search: besides the organization of the work of the analysts, the weekly meetings, and his own top search based on the SLT b-tagging algorithm, one of the things that occupied his days was the generation of large samples of VECBOS simulated events, which were then made available to the analysts for their studies of backgrounds. The work proceeded as follows. Claudio would define the number of events to be generated, along with the required input parameters, in a so-called "data cards" file. He would then submit the VECBOS job and the data cards to the batch queue of computers and wait for a week or so to get the output. The output was then passed to another program which turned into hadronic jets the list of quarks and gluons belonging to each produced event. It was a painstaking task, and it was made more complicated by the fact that often the jobs would crash after a few days of running.

One evening, Claudio went out in the company of Michelangelo Mangano for a quick pizza at Pal Joey's, a restaurant a few miles north of the lab which was very popular among Fermilab users. Waiting for their pizza, Claudio and Michelangelo enquired about each other's work.

> "And how is your simulation work going? Have you finally got the upper hand with VECBOS?" asked Michelangelo.
>
> "Oh, don't get me started with that. It's a nightmare, the code is not stable. I think there are some numerical exceptions that occasionally cause the jobs to crash." explained Claudio.
>
> "Ah, sure, I've heard about that. How many events do you run per job?"
>
> "I run 10,000-event simulations. It's the minimum, you know, to reduce the problem of large event weights ...."

Claudio was referring to an issue well known to Michelangelo. Whenever the simulated data were used to fill a histogram, one computed each bin content as the sum of weights of events falling in the bin; the resulting distributions correctly accounted for the varying probability of the simulated processes. The catch was that sometimes the simulation assigned very large weights to specific events, introducing huge uncertainties in the resulting histograms. Those large weights could only be dealt with by generating bigger event samples, where their effect would be mitigated.

> "... But they take about a week to complete on the batch system, and most submissions actually crash after two or three days. It's a painful bookkeeping job."
>
> "Bookkeeping? What do you mean?" asked Michelangelo.
>
> "Well, of course when the program crashes I collect the temporary output, the events that have been simulated up to that point, and store them, so at least I have not wasted the CPU. It's a pain, but piecemeal I am getting close to the goal of a million events with the standard choices for the minimum P_T and ...."

Claudio could not complete his sentence, as Michelangelo almost spilt the beer he was drinking on his freshly ironed Yves Saint Laurent shirt. His face went pale as he broke in:

> "Are you crazy? The jobs crash and you keep the output??"
>
> "... and Q-squared. Yes, what's the matter?" Claudio inquired.
>
> "Don't you know that VECBOS scans the parameter space progressively? If the jobs crash halfway, you are screwed! You have been collecting junk all along!"

Alas, Michelangelo was right: unlike other Monte Carlo programs, VECBOS performed an ordered scan. It was exactly as if you started to paint your house in the morning, went to lunch after covering 30% of the surface, and then came back to re-paint the exact same area in the afternoon: you could repeat this forever but you would never get the full job done. And neither could Claudio finish his pizza that evening: the thought of having to throw down the drain the product of months of work, spent joylessly wrestling with the Fermilab batch queue, was too much for his stomach. In his defence, we should note that data analysis tools in those years were not well developed, and they usually lacked a proper documentation. The instructions were passed from mouth to ear, and there was a general lack of sharing of resources and information. This was a real problem in the experiment, in fact, and one which would affect the coordination and synergy of the groups of analysts for years to come.

Enter Grassmann

Hans Grassmann was a German researcher. He was a strongly built, tall guy, who had soft manners when he was not discussing physics, but also the dangerous habit of speaking his mind regardless of the consequences. He had worked in the UA1 experiment during the years of the top-quark search and had made a few enemies then, since he had dared to criticize the alleged top signal, sparing no words as he thrashed the work of his colleagues. And he ended up meeting again some of them in CDF a few years later.

In 1989 Hans left UA1 and came to work for CDF in Pisa with a contract called "Articolo 22," a formula which allowed INFN institutions to hire foreign researchers for three years. It was Giorgio Bellettini, the head of the CDF-Pisa group, who had made the position available. Bellettini was tall and physically fit; he had been a strong tennis player in his youth, and you could still almost guess it by looking at his build and movements. He had been a schoolmate of Carlo Rubbia in his youth, but he had been the less lucky one in the hunt for a Nobel Prize. Bellettini now badly wanted to put together a strong analysis group in Pisa to search for the top quark, using the kinematic features of the signal rather than b-tagging to distinguish it from backgrounds. He understood that Grassmann was extremely intelligent and fast thinking, and he liked a lot the German's ideas on how a top-quark signal could quickly be evidenced in CDF data.

Bellettini also benefited of a constant flow of brilliant students who came to his office in Pisa asking for topics for their undergraduate theses or their Ph.D. To the students, the offer of working at the search for the sixth quark with the world's highest energy collider was irresistible. Since 1990, Bellettini could count on the help of two young and active graduate students: Sandra Leone, who had finished her Làurea thesis on a W asymmetry measurement in CDF based on Run 0 data, and Marina Cobal, who had moved to CDF after obtaining her Laurea on the Virgo experiment, a gravitational wave antenna. With Grassmann as a thesis advisor, the work could start on two fronts: Sandra would take on the search for dilepton top-pair decays; and Marina would search for top events in the lepton plus jets channel.

Bellettini had correctly understood that the kinematics of top-quark decay could be a performant weapon to separate top events from the

W+jets background. Hans and Marina devised an algorithm which relied on the energy of the jets to distinguish the signal from backgrounds: for each event, the algorithm computed a "relative likelihood" that the event was due to top-pair production as opposed to W+jets production. This approach was powerful, but it made the result very sensitive to the modeling of the tails of the distribution of jet energies. In other words, the likelihood strongly relied on how well VECBOS estimated the rate of high-energy jets. Grassmann knew that he was getting very close to the same trap into which the early UA1 top search had fallen: that reliance on Monte Carlo simulations was dangerous. The high-energy tails of jet energy distributions were a spot where VECBOS was largely still untested, and could give a wrong answer!

As I already observed, innovations are seldom greeted with the red carpet in scientific research. Much like Kondo's dynamical likelihood, the relative likelihood method only became an accepted tool in CDF several years after Grassmann left the experiment. Nowadays, the relative likelihood is routinely used as a *test statistic* (a function of the data suitable for extracting information from it) to exploit small differences in the separation of signals from backgrounds. Back then, though, it was looked with suspicion or even considered a faulty approach.

Using their kinematic selection, Marina and Hans could extract an intriguing excess of top-like events from the data. In December 1992 they published a CDF note which reported their search in the data collected during Run 0. They compared the distribution of relative likelihood in real data and VECBOS simulated events, showing that the data had a few outliers containing very energetic jets. Those events could be explained well by the production of heavy top-quark pairs.

The claims of the document were bold, and the analysis method quite in conflict with the conservative approach that the majority of members of the Heavy Flavor group wanted to follow. The mastermind of that conservativism was Alvin Tollestrup, and the gatekeepers were Campagnari and Yagil, who could count on the support of the leaders of a few American institutions eager to secure for their personnel a leading role in the top discovery. The consensus among them was that the top quark had to be discovered with a waterproof analysis, which could show that all of the predicted properties of top-quark decays checked out. This of course included the demonstration that the events contained two b-quark jets as demanded by the standard model decay of top quarks. The bare observation of kinematic characteristics in disagreement with Monte Carlo simulations was considered insufficient.

At the meetings of the Heavy Flavor group where Marina and Hans attempted to discuss the progress of their studies and defend their results, they found themselves facing an execution platoon, where hard questions got fired in rapid succession. The list of snipers was so long that one would not have been surprised to see a machine distributing numbered tickets at the entrance; Avi Yagil would have needed a stack for himself. In addition, the Pisa analysts had the feeling that cheap tricks were used to hinder the progress of their study. Marina's talks were often placed at the bottom of the meeting agenda, where the risk of getting rescheduled was very high. Marina and Hans were based in Pisa, and talks by video-conference were still very hard to deliver in those days, with the international connection often shaky and disturbed. The microphones in the conference room did not allow participants connected by video to hear the comments that participants on site would make from the back seats in the meeting room. Hence, Marina often flew in from Pisa in order to attend in person. Getting bumped to the next meeting two weeks later was a source of distress.

A witness that the agenda of the Heavy Flavor meeting was managed carefully in those years was once Paris Sphicas, who in 1992 had started to develop with the help of his Ph.D. student Baber Farhat an alternative to the SLT b-tagging algorithm of Campagnari and Yagil. Paris and Baber had found a way to improve the classification of electrons embedded in hadronic jets, by tightening the matching requirements between the charged track produced by the electron candidate and the energy deposit that the particle left in the CES detector. Paris once asked that his student's presentation be scheduled for the following week. A preliminary agenda was soon circulated to the recipients of the Heavy Flavor group mailing list, and Paris was happy to find out that Farhat would be the first to speak, before a presentation on the competing SLT algorithm by Claudio Campagnari. Talking first could be considered an advantage.

A little after receiving the preliminary agenda, Paris bumped into Avi outside of the CDF portakamps, in an area where smokers could entertain their vice without bothering their colleagues. It was late spring, and many offices in the trailers had open windows. As Paris and Avi chatted amicably, enjoying their cigarettes, they suddenly heard the unmistakable voice of Claudio. The co-convener of the Heavy Flavour group was talking on his office phone to a colleague, uttering profanities as he explained his disappointment to the person at the other end of the line:

"He sent around the agenda without asking me, and he put them first and me second!"

Avi awkwardly grinned to Paris: they both knew what that conversation meant. An hour later, a new meeting agenda was circulated: Farhat's name had slipped to second place. Fortunately, though, the four analysts later found a way to work together constructively on the SLT algorithm.

Don't Trust the Tails!

As I mentioned above, the criticism that Marina and Hans received meeting after meeting had strong roots in the alleged untrustworthiness of VECBOS in correctly simulating the tails of the jet energy distributions. This was an argument often made by Michelangelo Mangano, the in-house QCD authority. It caused real damage to the *Event Structure* analysis — that was the name of the top search of Grassmann and collaborators — as the theorist's voice was carefully listened to by his CDF colleagues. Michelangelo was objective in his assessment, but maybe he did not consider fully the consequences of his emphasis on the shortcomings of the simulation: his words created a lot more mistrust than could Marina's cross-checks create confidence. It was quite unfortunate, since to an experimentalist the good agreement between the tail of the jet energy distribution of real and simulated data that Marina could demonstrate in background-enriched samples should in general have been a sufficient validation. The mistrust in VECBOS caused increasing demands: extenuating lists of cross-checks were requested, and ideas of all kinds for further studies were put forth. Every time Hans and Marina resurfaced with material that addressed the objections and requests they had received in their past presentation, they got buried with new ones. Some were good suggestions meant to improve the robustness of their results and make them more trustable; others looked more like cunning traps engineered to slow them down.

Due to the extensive use of simulations, the focus on the tails of the kinematic distributions, and the vocal opposition of a few respected members, the consensus within the Heavy Flavor group was largely against the Event Structure analysis. But consensus is a volatile commodity, which needs to be attended to with care. Besides the conveners and the members of a few influential groups strongly polarized against the work of the Italians, the regular meeting attendees included a large number of silent observers, who used to sit in the back rows. Then there was a continuous influx of new members, as was normal in an expanding collaboration. A constant evangelical effort was needed to keep educating the audience on the dangers of trusting background simulations. One clear example of that effort is the presentation which Paris Sphicas was once asked to give. This had to focus on UA1 studies of the sample of W+jets events isolated in that past experiment at CERN.

The idea came one morning to Claudio Campagnari and Paul Tipton, when they went for a cup of coffee at the third floor of the B0 building. They started chatting about the discussions that had taken place the previous afternoon at the Heavy Flavor meeting. As Claudio prepared a coffee, they were joined by Steve Geer and Paris Sphicas. Sphicas and Geer had been UA1 members before joining CDF, and there they had studied W+jets events carefully. In fact, UA1 had famously puzzled for a long time over a couple of events in the tail of the W p_T distribution.

When a W boson is produced in a hadron–hadron collision, it normally carries little momentum in the direction transverse to the beams. The initial transverse momentum of the colliding quarks is quite small, and most of their energy is spent to create the massive boson. However, sometimes, the collision releases extra energy, which is emitted in the form of QCD radiation (e.g., an energetic gluon). In this case, the radiation recoils against the W, which may consequently acquire a significant amount of transverse momentum. Experimentalists measure the latter using the observed momentum of the charged lepton and the missing transverse energy produced by the W decay. If one uses VECBOS to predict the transverse momentum in W boson candidates, one obtains a distribution peaking at 10 GeV or so, with a long tail extending to higher momenta. However, in the days of UA1 nothing like VECBOS was available to predict that distribution. Physicists relied on a more generic simulation, which could handle the production of a W boson recoiling against one single quark or gluon. Additional gluons had to be generated by artificially modeling QCD radiation, with large resulting uncertainties due to the scarce knowledge of those processes.

The two outlier events observed in UA1 data were far away from the rest of the data and from the unrefined background predictions available at the time. Carlo Rubbia and Steve Geer had thus hypothesized that they were due to a new physics process. If a new heavy resonance were produced and decayed to a pair of W bosons, with one of them yielding a lepton–neutrino pair and the other a jet pair, this could very well explain the two odd outliers. Sphicas, who had done his Ph.D. thesis on the search for hadronically decaying resonances in UA1, had studied those events in detail. He was, therefore, well aware of how the shortcomings of a background simulation could lead to interpret few events in the tail of a distribution as due to some new physics processes, when they were in fact due to ill-predicted backgrounds.

The story of the UA1 analysis in the W+jets sample was well-known to Claudio and Paul. So that morning they asked Paris to prepare a presentation for the next Heavy Flavor meeting, where he would show the pitfalls of relying on the tails of Monte Carlo simulations, and the troubles that a past experiment had gotten into by doing that. Paris agreed, and on the following Thursday, after all the subgroup reports, he got the floor. He showed the UA1 dataset and how those two events stood out, unexplained by the almost vanishing background predictions in the W p_T distribution. Then he proceeded to tell the story of how the simulation had been underestimating the rate of W plus two jet events, leading the experiment to put forth some claims that it would later have to retract. The W p_T distribution shown by Paris was qualitatively identical to the jet E_T distributions that Marina and Hans had been showing since 1991: a rapidly falling spectrum, a void, and then some outliers. Top events? We now know they quite possibly were, but back then they had to be dismissed as an ill-understood background contamination. Avi Yagil was beaming; he and a few others ostentatiously started clapping hands the moment Paris finished his talk.

To make matters worse for the Pisa analysis, some of Grassmann's colleagues felt he treated them with contempt, as if he were affected by a superiority complex. They claimed he was not collaborative; perhaps, the perception was due to his attempt to defend himself from the too many punches below the belt he was receiving in the Heavy Flavor group. He even suffered nasty jokes, like a schoolboy targeted by his peers. A young Italian colleague once malignantly wrote and launched a script that issued a continuous loop of "tape mount CCB112" and "tape unmount CCB112" commands to the vault system where the data tape was stored. The sequence continued to take in and out of the reader the specified data tape, with the effect of inhibiting anybody from accessing its contents. That was the tape which Grassmann was trying to spin in order to perform a requested cross-check to his analysis! Similar obnoxious acts of sabotage were fortunately rare. One thing was the clash of personalities, and quite a different one was the understanding of physical reality that every soul in the Heavy Flavor group would have sworn to be the ultimate goal of all his or her actions.

Objectively speaking, the b-tagging-based analyses of the single-lepton final state were not much more effective in unearthing top-pair events than the Event Shape analysis, but they were designed to be more robust. They also had an additional advantage: the b-quark tag made it easier to reconstruct the top-quark mass in the selected events, as it reduced the ambiguity on the choice of which of the four jets were to be assigned to b-quarks and which to light quarks from W decay.

The reconstruction of the most likely value of the top-quark mass was a crucial part of the whole search. Once an excess of top-like events were identified by a *counting experiment* — one where the number of collected event candidates is compared to background predictions to make inferences on the presence of signal in the data — CDF would have to prove that these really behaved as expected from top decays; that included showing that they were compatible with a top quark of the same mass. Because of that, a large part of the activity of analysts involved in the top search had the goal of enabling a precise reconstruction of the top mass, something which could be done with a kinematic fitter.

Several groups had independently attempted to put together a performant kinematic fitter. Claudio and Avi had also tried once, getting to the point of convincing themselves that it was a lost cause given the small number of reconstructable top events they could expect to collect. Claudio remembers it as one of the worst mistakes of his whole career. Alvin Tollestrup attempted to convince Claudio that even with few events a mass fit was worth trying, but Claudio did not really take the suggestion seriously. He was deeply convinced that it would be impossible to measure the top mass with only a handful of single-lepton top candidates. He thus let slip from his hands the satisfaction of being among the first to see the top mass peak appear in the data. The two groups that pulled it off were the ones from Pennsylvania and Berkeley.

The Pennsylvania group advocated the use of a kinematic fitter based on Fred James' MINUIT program, a cherished numerical algorithm that could be used in any minimization problem. Instead, the Berkeley group had been developing a fitting method based on a more specialized program called SQUAW. Both methods used the decay chain that the standard model predicted for top-quark pairs. They subjected the measured jets and leptons momenta to constraints consistent with the assumed kinematics, in order to extract the few unknown quantities in each event. The constraints were of three kinds: the compatibility of the combined mass of jet pairs and lepton–neutrino pairs originated from W decays with the known W mass value, the equivalence of the top and antitop masses, and the global momentum conservation laws that the event had to oblige to. The unknowns were the z-component of the neutrino momentum, which was not measurable by the apparatus, plus of course the most likely value of the top-quark mass. However, while MINUIT directly constructed the function to be minimized from the constraints (a global *chi-squared*), SQUAW considered the decay chain from the point of view of the subsequent decay vertices of top quarks and W bosons, and used the technique called *Lagrange multipliers* to enforce the experimental constraints. After many discussions and performance comparisons, the fitter of the Berkeley group was finally chosen. It was a tough decision, which threw away months of work of the Pennsylvania scientists; but a choice was necessary.

Disclaimer: No Top

Until discussions were held at the Heavy Flavor meeting or in other internal arenas, the periodic rise of tempers and the occasional name calling could be seen as an acceptable collateral damage to take for the sake of finding a consensus, or at least a majority view on how to progress with the science. The arguments were often bitter, but the subject of the discussion was always a concrete one: what algorithm to choose, whether a statistical method was sound or not, what cross-checks would be required to make a graph publishable. CDF thrived despite those hot internal debates. The enormous wealth of the collaboration was the competence, the knowledge, and the experience of its members; and the best way to use such assets was through a free communication. The collaborators spoke frankly, and different ideas were compared in order to find the best one. The shortcomings of flawed methods were mercilessly exposed regardless of the pain this sometimes caused.

The matter was entirely different if discussions arose in a public arena, such as occasionally happened at international physics conferences. In that case, they could shatter the image of unity that the collaboration wanted to show to the outside world. Open arguments could give the impression to outsiders that the results published by the experiment were controversial and did not represent the consensus of the collaboration. In short, they put in danger the authoritativeness of the scientific output of CDF.

At regular intervals, the collaboration decided what material to present at international conferences. In order for a result to be accepted for free distribution, it required a formal "blessing" at the working group of relevance. Ultimately, the decision on whether the result was blessed rested in the hands of the group conveners. Responding to the insistence of Grassmann, who wanted his Event Structure analysis of Run 0 data blessed such that the results could be shown at the important Moriond 1993 winter conference, Campagnari said adamantly what he thought: "Until I am alive, this stuff will not make it out of here." Campagnari's shocking candour caused outrage in the Pisa group, but he had his back well covered. The large majority of CDF agreed that the experiment was not ready to present those results, as they sounded as a claim that the W+jets data showed a signal due to top-pair production.

A few months later, the issue of what to allow for outside distribution came up again, as a very important international conference was scheduled in Marseille at the end of July 1993. The laboratory director John Peoples had been invited to give an overview talk, which included top search results from both CDF and DZERO, but he declined and suggested that Lina Galtieri substitute him; Lina accepted to carry out that important task. She knew that a few top candidates had been recently spotted in b-tagged single-lepton categories. Before leaving to France, she in fact spent her time with Weiming Yao to reconstruct seven beautiful events that her Berkeley collaborator had selected, rather than writing her talk as she had planned. Lina enjoyed the complete trust of CDF; a leak of internal information during her talk was unconceivable.

The real question was whether anything about the Event Structure could also be said at the conference. Bellettini this time had a convincing argument in favor of that. His student Marina Cobal was now close to finishing her Ph.D. thesis. A physicist who could not demonstrate the trust of her colleagues by speaking on behalf of the collaboration would be seriously hampered in her post-Ph.D. job search. The chance to present CDF results would repeat itself after Marseille, yet the problem would not go away. On the contrary, those events in the tail of the jet E_T spectra had started to be regarded, for good or bad, as a systematic effect rather than a statistical fluke. Hence, it was only natural to expect that additional events would be found with the addition of new data that the Tevatron was continuing to deliver. Furthermore, a heavy top quark — one with a mass in the 150–170-GeV range — was now being suggested by indirect information collected by the LEP experiments. This made Grassmann's analysis a problem destined to grow bigger for CDF, as its sensitivity to the top signal was comparatively higher if the mass of the particle was larger.

A compromise had to be struck. The conveners discussed the matter with the spokespersons, and a solution was found. Marina would present a short talk on results of "New particle searches at CDF" in a parallel session, and in her slides she would be allowed to insert one graph from her own analysis. The graph showed the characteristics of her top candidates in the plane defined by the measured transverse energy of the second and third jet. That "$E_{T,2}$ vs $E_{T,3}$" graph was the "money plot" summarizing the Event Structure search. The figure compared the VECBOS expectation to real data that had been collected until then during Run 1. In showing the graph, it was agreed that Marina would read a carefully crafted sentence, a disclaimer stressing that CDF did not consider the outlier top-like events an evidence of top production in their data, due to the untrustworthiness of the background distribution they were compared with.

Marseille 1993 turned out to be a huge conference for particle physics standards: 800 participants from 39 countries around the world gathered there at the end of July. The conference was organized in two three-day parts: an initial phase of 19 parallel sessions, where short presentations could be made on very specific topics, followed by a second phase of plenary sessions with broader summary talks. This was an important time for particle physics. The LEP experiments had recently collected a much larger bounty of Z bosons than what had been accumulated in previous years, thanks to the improved performance of the accelerator. There were obviously great expectations for the precision measurements of electroweak parameters that they would present. In addition, of course, many attendees were curious to know whether the Tevatron would show first indications of a heavy top quark.

Marina gives her talk in parallel session 9, in front of scores of colleagues interested in new particle searches, the topic of the session. This is her first important talk at an international conference, and she is visibly tense and a bit scared. Finally, she places on the projector the controversial transparency with the $E_{T,2}$ vs $E_{T,3}$ scatterplot, with the tantalizing clustering of three black data points on the top right corner, where backgrounds can hardly contribute. Facing the large screen where the slide is projected, she waves at the graph with a pointing stick.

> "So as you see in this plot, the black points are experimental data from Run 1, and the boxes indicate the expected contribution from W + jet backgrounds as predicted by the VECBOS simulation."

Then she duly supplements the description with the mandatory disclaimer, reading aloud the text spelt in thick black letters under the graph:

> "Due to the large uncertainty in the modeling of backgrounds, this distribution cannot be taken as an indication of a departure of the data from known sources."

But then Marina turns her back to the screen and faces the audience, looking at them as she adds, with a different tone of voice:

> "I have been asked by the CDF collaboration to read you that sentence, however I believe that these outlier events are hard to explain with backgrounds alone, while they fit quite well with the hypothesis of being due to the decay of pairs of heavy top quarks."

The audience is taken aback. CDF has a top signal after all?! Among the listeners are several of Marina's CDF colleagues. One of them is Melissa Franklin. She stands up and publically objects, criticizing with colorful words the bad service that Marina is doing to the collaboration. But Marina is not intimidated and retorts that as a scientist she has the right to express her personal views. After some further exchanges, the convener manages to stop the discussion, but the scientists in the room have gotten the message. CDF may be seeing the first top events, and there are internal conflicts on how to handle the matter. But that is only the beginning: in the course of the following months, the rumors about the status of top searches in CDF will continue to grow.

Chapter 7

The Discovery of the Top Quark

Toward the end of the summer 1993 CDF went undercover. The spokespersons Shochet and Carithers decided that the evidence for top production in the data was mounting; It looked likely that soon they would be able to make the public announcement of seeing the long-sought sixth quark. Results were feverishly expected from the analyses of the full dataset of 20 inverse picobarns of collisions that CDF had collected in the just ended Run 1A. It was considered dangerous for the experiment to produce interim, inconclusive results which could backfire if unsupported by future data.

"Never Give Talk After Brian Greene"

The turning point happens shortly before the XVI Lepton–Photon conference, which is scheduled to take place at the Cornell Laboratories on August 10. During a Top group meeting, Mel Shochet and Alvin Tollestrup are taking notes of the event candidates and background predictions of the different counting experiments, as they hear the subgroup leaders report the status of their searches for the top quark in the various channels. Dileptons, two events observed, 0.56 predicted; SLT single leptons, seven observed, 3.1 predicted; SVX b-tags, six observed, 2.3 predicted; and so on. At some point, Mel and Alvin look at each other: taken singly, none of those results is significant, but together they start to be! A lively discussion ensues. The decision is finally taken to stop presenting updated search results at conferences until a definite, well-supported claim can be put forth.

As the meeting ends, Claudio walks back from the Hirise to the trailers together with some colleagues. It is a sunny summer evening, and it is quite pleasant to walk rather than drive the few hundred meters that separate CDF from the Wilson Hall. The only concern is to avoid the droppings of geese, the unchallenged masters of the grass and pathways throughout the lab. Everybody is cheerful: they finally have something in their hands! The top discovery is close. Spirits are high, but as the CDF collaborators near the B0 parking lot they run into Paul Tipton, who has just parked his car in front of the trailers. His face is livid. He is evidently upset.

> "What's the matter Paul? Aren't you happy about the great results we've seen today?!" asks Claudio.
> "Goddamnit, I missed it by one! I missed it by one!" is Paul's answer.
> "What do you mean?" Claudio inquires.
> "I volunteered for the talk at Lepton–Photon several months ago, as I had calculated that by the time of that conference we would have an evidence of top in our hands! I wanted to be the one who would show it, and I missed it by one!"

A smile spreads out on Claudio's jovial face. Paul and he are used to tease each other at any possible occasion, and this is not one he wants to miss. So he replies:

> "Oh! That's right. Well, I wouldn't bother going: from what I've heard at the meeting you will be showing the same stuff I've showed in my talk at La Thuile four months ago — and there's no skiing involved in Cornell. You can bring your skates, though."

Paul is not amused as he explains:

> "I *am* going — I have already bought my flight."
> "Well, okay. In that case feel free to re-use the slides of my La Thuile!" is Claudio's sarcastic reply.

Things indeed turned out badly for Tipton. He was not able to give a *lectio magistralis* on how he had discovered the top quark; that would be somebody else's job at a later occasion. His talk was devoid of any new information on the CDF top-quark search. He also had to watch his words carefully, since there was a significant gap between what he could say and what he did know about the top search. Furthermore, Tipton suffered a disappointment of a different nature during his session: the speaker who preceded him was Brian Greene, one of the most influential theorists around. The conference hall was packed full as everybody wanted to hear Greene talk; his contribution was titled "What can we do with String Theory?" At the end of his talk, Greene lingered on the stage to answer a long series of questions from the audience, leaving Tipton impatiently waiting for his turn to speak. As Tipton finally took the podium and started fiddling with the microphone, he heard the unwelcome sucking sound of the conference hall suddenly getting half-emptied by attendees leaving the room. He recalls his thought: "Important note to self — never give talk after Brian Greene!"

As for the conference scheduling, it later transpired that Tipton had not missed it by one; he had missed it by several, since it would take more than seven additional months before CDF would be ready to talk about new top-quark results in public arenas. Those months were spent finalizing the analyses of the full Run 1A dataset, performing dozens of cross-checks, and slowly building a consensus around an unavoidable conclusion. The various excesses of events in the different search categories, the kinematic evidence coming from at least three independent analyses, and the tantalizing mass distribution of single-lepton event candidates could all be explained by a heavy top quark, and only by it.

The question then became what was the right way to publish the results. The original plan of the management had been to produce one short paper for each of the various independent analyses, and a further summary paper wrapping it all up and including the result of the mass measurement analysis. The four articles (one about JETVTX tags, one about SLT tags, one about dilepton candidates, and the summary one) were meant for publication in PRL, a journal which had a strict limit on articles length. That was a problem: by the time one was done with the mandatory introduction (i.e., why it is important to search for a new quark, what theoretical indicia point at its existence, what has been seen in past searches, etc.), the space left for a detailed explanation of the experimental technique, the cross-checks, the findings, and the interpretation was minimal. Alternatively, one could think of writing four inter-dependent papers, but that would run into the opposition of the journal. The short PRL papers were drafted between September and October 1993 by the researchers who had performed the different analyses. Then they were submitted to the collaboration for an internal review. Unfortunately, nobody liked them: taken one at a time they were very inconclusive. Some collaborators even let go with adamant remarks; one of the drafts was openly called "pathetic" by the leader of an important group. This of course did not help, as it created an even tenser situation.

To the impasse of four inconclusive papers, one should add that Bellettini had written a fifth PRL draft describing the Event Structure analysis; that paper did not belong to the original plan, but it now was there and had to be reckoned with. After years of fights, the Event Structure had come to be perceived by many as an honest analysis, well-checked, and worth complementing the counting experiments and the mass measurement. Furthermore, the events that the analysis flagged as top candidates largely coincided with the top candidates selected by the more standard b-tag analysis. Yet, the basic disagreement on the reliance on simulations had not been resolved, and the core group of opposers had not changed position. The collaboration was still split on the matter.

The sociological repercussions of the tense situation were hard to manage. The original plan of having four PRL articles had the benefit of administering to the main authors of the different analyses the honor of being editors. It was now difficult for the spokespersons to take that honor back. Yet, there was no other way: making a strong case for the top-quark evidence was too important a goal, one which could withstand no compromises. The short papers eventually got dumped, and Shochet and Carithers opted for a big article to be published in *Physics Review D* (PRD), a journal which did not enforce any page limit. The spokespersons designated Tony Liss and Paul Tipton as editors of the PRD article.

While Tony and Paul were busy writing the text, leaving blanks where missing information still needed to be inserted, all analyses continued to polish off their results and produce supporting material. The article was also agreed to contain a chapter on the Event Structure analysis, and Bellettini set out to write it. Unfortunately, his draft was heavily shortened in the final version, which completely avoided any statement about the *statistical significance* of the distributions of relative likelihood observed in the data.

What is Statistical Significance?

As we near the discussion of the discovery of the top quark, we need to make a digression to explain an important concept used by particle physicists to measure the level of surprise of an observation, i.e., how much are data at odds with a hypothesis. In a nutshell, the statistical significance of an observed effect (usually expressed as a "number of sigma") is a function of the estimated probability to see an effect at least as surprising in the data. Z-values (as significances are sometimes called) and p-values (as statisticians name probabilities) are thus connected by a mathematical map, a one-to-one correspondence. As you are more familiar with the concept of probability, the way that map is constructed can best be explained by starting with an example involving probability estimates.

You and your friend Peter go to a Casino one evening, and quickly lose track of one another, as you join a table of Craps, while he moves to a Roulette table. Your game that evening turns out to be a quite extraordinary one, as at some point you throw 10 "naturals" in a row. At the game of Craps, a natural occurs when the two dice show a combination among the set (1,6), (2,5), (3,4), or (5,6). A streak of 10 naturals is *really* uncommon, and as you later start looking for Peter, you cannot wait to report your feat. When you finally find him, Peter looks even more excited than you, and he is the first to explain what happened to him: at his Roulette table, he observed a sequence of 20 *rouges* (omitting the times when the neutral "zero" came out). Once you also report your experience, a nagging question unavoidably arises: should the sequence he witnessed be regarded as more, equally, or rather less surprising than yours?

Admittedly, the surprise caused by an empirical observation can often be quite subjective. If, on the next day, you and Peter join a group of friends and tell them your Casino adventures, their answers to the question on the relative surprise level of the two observations might come in roughly equal proportions. Yet, there is of course only one "right" answer; the problem is mathematically well defined and can be solved by numerically comparing the probabilities of the two sequences of events. Since the throwing of dice at the Craps table and the landing of the ball on the Roulette wheel may be supposed to be random processes, and each observed outcome should be independent of all others, then the probability of sequences of events is obtained in both cases as a simple product of relative frequencies. A natural occurs on average twice every nine trials, so the probability of the sequence of 10 naturals must be (2/9) to the 10th power, which is three-tenths of a millionth. For the Roulette sequence, the probability is instead (1/2) to the 20th power, which is just a bit less than one-millionth. Both series are thus quite unlikely, but in quantitative terms the sequence of dice throws is three times more unlikely than the sequence of *rouges*. If we were Casino managers in search of indicia of fraud, we would have to consider the dice throws a stronger *notitia criminis* of something awry than the Roulette results.

The above example indicates that very uncommon sequences of events force us to deal with *very small p*-value estimates. Now, there is no problem when one writes a small number in a formula: scientific notation comes to the rescue, and powers of 10 elegantly substitute cluttering strings of zeros. So we succinctly write 3×10^{-7} in place of 0.0000003, for instance. Still, when one *talks* of a small number, one prefers to avoid spelling out those exponents. When they have to deal with very large or very small numbers, scientists in fact introduce multiples or submultiples: this usually calls for the use of the suffixes kilo-, Mega-, Giga-, Tera- or milli-, micro-, nano-, pico-, and so on. For *p*-values, however, the standard conversion does not involve powers of 10, but a slightly more complicated recipe based on the size of the area under the tail of a *Gaussian function*.

The Gaussian function, also called "bell curve," is a cornerstone in statistics theory: it describes the typical probability distribution of measurement uncertainties. If you measure a physical quantity with some instrument, the values you obtain by repeating the measurement usually scatter around the true value by following a Gaussian probability distribution, which characteristically predicts that larger and larger deviations are increasingly less frequent. The width of that Gaussian curve is proportional to the parameter called "sigma," indicated by the Greek letter σ. For a Gaussian measurement, the sigma parameter coincides with the *standard deviation*; a smaller sigma corresponds to a more precise instrument, i.e., one which returns an answer within a smaller interval. In fact, the standard deviation of a set of data is a measure of their "typical" spread around their mean.

The conversion of a probability into the more manageable number of standard deviations — i.e., sigma units, or Z-values if you prefer — is performed by finding how far out in the tail of a Gaussian distribution, in units of sigma, one has to go in order that the area of the distribution from that point to infinity be equal to the stated probability. As a result of this procedure, a 15.9% probability can be seen to be equivalent to a one-standard-deviation effect (boh-ring!); a 0.13% probability is a three-standard-deviation effect (hmm, exciting!); and a 0.000029% probability corresponds to five standard deviations, "five sigma" (woo-hoo! a discovery!). Those numbers no doubt appear meaningless to you at first sight; yet, once you get familiar with the above correspondence, you will find it easy to assess statistical effects using Z-values. To see how, let me make one further example.

When you report a deviation of an estimate from the reference value of a quantity you are measuring, you are accustomed to employ a unit appropriate to the measurement you are performing. So, for instance, for your body weight, you could use the kilogram. Upon stepping on a scale you might utter: "I expected to weight 80 kg, but according to this goddamn thing it's 82!" Note how this sentence is not very informative from a scientific standpoint, as it says nothing about the intrinsic precision of the instrument. Because of that, we cannot conclude that you have gained weight, as that conclusion depends on whether your scale measures weights with a standard deviation of, say, 2 kg, or rather 0.5 kg. In the former case, there is a sizable chance that your weight is actually 80 kg and the scale happened to report a measurement one standard deviation higher than the true value; while in the latter case, you have almost certainly been delusional about your body weight.

A different way to report your measurement, one immediately providing information on the significance of the discrepancy, is to express the deviation as a Z-value. This is counted in units of sigma of the Gaussian distribution which corresponds to the precision of the instrument. Again, if sigma were 2 kg, which is exactly the difference between 82 and 80 kg, then Z would equal 1 and you would coolly say "My weight measurement today returned a value one sigma higher than I expected," which a statistics-savvy listener would not take as evidence for a weight gain. In fact, a statistical fluctuation of the measurement is a reasonable explanation of the discrepancy: a one-sigma or larger departure (in either direction) happens by mere chance 31.7% of the time, without the need for the true value of the measured quantity to have varied. If sigma equaled instead 0.5 kg, then 82 and 80 kg would be *four* sigma apart, so Z would equal 4; this could now be taken as strong indication that you have gained weight, as a four-sigma or larger fluctuation happens only 0.0063% of the time for a Gaussian measurement. I hope the above example allows you to appreciate the usefulness of relating deviations of a measurement from an expected value in sigma units.

Armed with the understanding of the concept of statistical significance, we may now return to the race to the discovery of the top quark.

Evidence!

One of the crucial issues that remained to be addressed was how to compute the combined statistical significance of the three counting experiments (the JETVTX and SLT analyses in the single-lepton channel, and the dilepton search). Taken together, the numerical excesses were marginally significant in terms of selected events over expected backgrounds, but appreciably more so in terms of the total number of b-tagged jets that those events featured. The decay of top-quark pairs always yielded *two* b-quark jets, while b-tagged jets in W+jets background events were mostly spurious: there, the presence of one b-tag did not significantly increase the chance to find a second one in the same event. This implied that one could expect to obtain a more significant excess from the top signal if one counted the total number of observed b-tags in the selected b-tagged events, rather than simply the number of b-tagged events. The counting of b-tags was justified by the more complete use of the available information (the presence of more than one jet with a b-tag). That choice appeared to some researchers an unjustified change of method, one adopted *a posteriori*; others, however, considered it a quite reasonable and defensible choice. It was not without some bloodshed that a consensus was finally reached. The paper would quote the total number of b-tags as the main metric on which to judge the size of the top-like excess above backgrounds.

The combined significance of the numerical excesses over background predictions computed with Method one amounted to 2.8 standard deviations. That was weak evidence of the new particle, and certainly not an "observation-level" effect. However, there was additional and independent evidence for top in the distribution of invariant masses reconstructed with the SQUAW fitter, which peaked at a value far higher than those where backgrounds were expected. It was also encouraging that a fit to that mass distribution yielded for the top mass the value of 174 GeV. This was in excellent agreement with the value suggested by the latest electroweak fits obtained by the LEP experiments, when the whole set of measurements produced on Z boson characteristics was interpreted in the context of a standard model with six quarks.

Despite the beautiful result of the mass analysis, the CDF members could not agree to include that information in a global significance calculation. A combined "counting plus mass" significance would risk to be perceived as far-fetched by the scientific community; and in any case, the result would not pass the threshold of "five-sigma" deemed necessary to claim a definitive top-quark discovery by the particle physics community. The collaboration had spent several months to converge to a solid consensus on the fact that the weak signals collected by the different counting experiments together made the case for a result worth publishing as an evidence for top. The added kinematic evidence coming from the top mass peak and the Event Structure analysis had been instrumental in aggregating that consensus, but they were not used directly in the paper.

The arguments on how to call the signal in English in the title and abstract of the article were also quite hard-felt. The word "evidence" was not so common in particle physics publications at the time, nor was its semantic equivalence to a "three-sigma" effect as well-known as it is nowadays. At one of the daily meetings held in the 1-West conference room of the Hirise Jacobo Konigsberg, a young post-doc from Harvard University who participated to the dilepton top search, argued: "If we call it evidence, doesn't it mean that we are saying it is evident that we have found the top?" Others wanted to remove any qualification from the sensitive spots in the article (title, abstract, conclusions), to stick to the numbers and phrase the title as "Search of Top-Pair Production."

Of course, conservativism was like a religion in CDF in those years. The top seekers had been really careful in assessing the significance of their result and had done their homework tidily. Even the toughest skeptics among their colleagues were finally convinced that the observed effect could have no other reasonable explanation than the pair production of heavy particles which decayed to W bosons and bottom quarks. Not calling them top quarks would have been close to obstinacy. Yet, many dragged their feet. It was surprising to see that old-schoolers such as Tollestrup, Shochet, and Bellettini were now willing to commit to stronger statements in the abstract, whereas it was the young guys who stubbornly played it down. Campagnari insisted that there were things that did not add up well. First of all, in the control sample of Z+jets events, there were two nagging events with secondary vertex b-tags, which could have nothing to do with top quarks (the top never produces a Z in its decay) but constituted an unexplained excess qualitatively not dissimilar to the one claimed to come from top production. And then there was a problem in the subset of W+4 jets events: the large cross section of top-pair production which could be measured in that subset alone was slightly at odds with the overall interpretation of the data.

In the end, Claudio, with the help of his friend and "conservativeness jihadist" Avi, managed to convince his colleagues that a disclaimer had to be inserted in the abstract. The term "jihadist" is from Avi himself: "We were jihadists, but we were right!" he claims nowadays, and his postmortem assessment is basically correct, although I would invert the order: they were right, but they were jihadists. They had realized that particle discoveries needed to rely on data-driven methods, hence they could not bring themselves to trust Monte Carlo simulations in any way. From there on, what they had been fighting was a war of religion.

After careful polishing, the long paper looked ready to be released. On April 26, 1994 CDF submitted to *Physics Review D* a bulky 152-page-long article titled "Evidence for top quark production in proton–antiproton collisions at sqrt(s) = 1.8 TeV." Once typeset in PRD two-column format, it fit into 60 pages of the journal. The abstract of the paper itself was 20-lines long. It included the following crucial statement:

> "The probability that the observed yield is consistent with the background is estimated to be 0.26%. The statistics are too limited to firmly establish the existence of the top quark, however a natural interpretation of the excess is that it is due to ttbar production."

As you can see, there was no explicit claim of a discovery in the abstract. Yet, the "natural interpretation" was put forth in clear. Further, mass and cross-section measurements were offered in the latter part of the abstract rather than only discussed in the body of the article: this by itself was a very strong message. It transpired that those were trusted to be real properties of the new particle; the CDF physicists attached sufficient meaning to those numbers to place them in the front page.

The statement that followed the one quoted above in the abstract was one that Claudio Campagnari had struggled to insert:

"We present several cross-checks. Some support this hypothesis, some do not."

That was indeed a sobering remark!

Simultaneously with the submission of the "Evidence PRD," as the paper came to be dubbed, CDF gave a seminar to explain the result to the Fermilab community and to the press. And, on the same day, Bellettini's group published a public INFN note where they described the results of the kinematic analysis. This unusual move had been agreed in advance with Shochet and Carithers, to allow the Italian physicists to get their share of spotlights, at least in Italy.

The reaction of the world of particle physics to the claim put forth by CDF was a mixture of enthusiasm and skepticism. For sure, the top-quark events that CDF was presenting had all the required characteristics to be a genuine signal. There was a numerical excess of events with b-tags, passing a kinematic selection believed to enhance the signal fraction. And there was a remarkable mass distribution, where six out of seven events piled up at the same place in a histogram which looked too good to be true. On the down side, the significance of the excess of data was not overwhelming, and it was sewn together by combining small effects which individually meant little. In addition, there were all the subtle issues on which Campagnari and the "young guard" had based their request for a low-lying announcement.

The first important international conference after the publication of the CDF evidence paper was ICHEP 1994, which was held in Glasgow at the end of July. Three presentations of the CDF top searches were given there by Hans Jensen, Franco Bedeschi, and Brig Williams. Jensen's was a plenary talk and it summarized all the results contained in the "top evidence" PRD. Bedeschi and Williams instead presented parallel session talks. While Bedeschi focused on the counting experiment results, Williams explained the top mass measurement together with the Event Shape analysis of Grassmann and Cobal. Now that the b-tagging results were out, it was possible to discuss that alternative search method and defend it publically. Indeed, Williams spent the largest part of his presentation discussing the kinematic analysis, the relative likelihood method, and the detailed comparisons with VECBOS in the signal-depleted control sample of W+2 jet events. To insiders aware of the endless struggles that had taken place in the Heavy Flavor group, it looked surprising that the leader of one of the institutions which had initially most objected to Grassmann's analysis would present it in detail at an international conference. But CDF had proven that it could handle the internal controversies and finally digest them, turning them into great science.

Run 1B

Run 1B (1994–1995) had started in the meantime. CDF now featured a new silicon vertex detector, SVX' (read: svx-prime). It was built in close resemblance to its predecessor, but was crucially equipped with radiation-hard readout chips that could withstand the harsh environment of the center of CDF without a significant deterioration of their performance. The Tevatron soon began breaking one instantaneous luminosity record after another, and good data piled up at an accelerating pace. It would only be a matter of time before one of the two experiments reached a five-standard-deviations excess in one of the top-search channels and decided to finish the business, putting forth an observation claim of the new particle. The intrinsic randomness of particle collisions could play tricks, as small statistical fluctuations could decisively favor one experiment with respect to the other. For instance, a couple of extra dilepton events could tilt the balance. Note that the dilepton final state was one where the two experiments had roughly equal sensitivity, as no secondary vertex b-tags were required in the event selection there.

In the single-lepton searches, due to the lack of a microvertex detector, DZERO could only tag b-quark jets "à la SLT," by reconstructing low-momentum muons contained within hadronic jets; hence, it was at a marked disadvantage vs CDF. On the other hand, while CDF had refused to use the "Event Shape" information for significance calculations, DZERO could, in principle, decide to exploit something similar to overtake its rival. And indeed, DZERO was using two kinematic variables constructed out of jet energies to enhance the top content of their selected data. Using Run 1A data, the competitors had published a slight excess of events (9, when 3.8 were expected), as well as a spectacular dilepton candidate which had for a while brought the experiment to caress the idea of publishing a top-discovery claim based solely on that event. In the end, DZERO had only produced a 131-GeV lower limit on the mass of the quark, and an excess of 1.9 standard deviations over expected backgrounds; but, in their paper, they noted that their sensitivity to the top quark was similar to that of CDF. So, now the tension was breaking a record high across the ring, and the idea that DZERO could grab the prize of a definitive top discovery in the final rush, after the feet-on-the-ground decision of CDF to only put forth a solid, well justified and unassailable claim of three-sigma evidence a few months before, was nothing short of a nightmare to the CDF members.

Of course, with two experiments working only a mile apart, whose members shared lunch tables in the cafeteria of Wilson Hall, there was a confidentiality issue. CDF and DZERO members who worked at the top search were under big pressure, and lunch time was one of the few moments of their daily routine when they could exchange a relaxed word with colleagues about their work. This entailed the danger of involuntarily leaking information to the members of the competing collaboration, who might be sitting a few feet away. The researchers were careful: it was against their interest to inform the competition on the status of the ongoing analyses. But despite the care, words spread quickly in the lab. Fermilab theorists, who often had lunch with experimentalists from the two collaborations, craved for any hint about how the searches were going. They sometimes ended up acting as messengers between the two otherwise non-interacting groups, as they traded in small bits of information acquired from chats with DZERO when they asked for hints from CDF, and *vice versa*.

Among those most worried by the possibility of leaks was Alvin Tollestrup. His office was close to the one of Brenna Flaugher, who was married to Tom Diehl, a DZERO member. Crossing Brenna in the hall every day kept reminding Alvin of the impending danger. One day, he could not hide his worries with her.

> "Brenna, there's one thing I would like to ask you — you know, your husband works in DZERO, and I am wondering whether you are paying enough attention to avoid giving out information on the status of our top searches ...."
>
> "Oh, we never discuss internal information among us, we have much more fun talking about the people. And besides, you can find out more at lunch tables in the cafeteria than by direct contacts with the other collaboration's members."
>
> "Are you sure? I mean, you never let go any secrets about our analyses? I might imagine that during bedtime chats, for instance ...."

Brenna was so surprised by Alvin's insistence that she figured he must be kidding, so she decided to play along, by telling a lie so blatantly false that the argument would be forcefully closed.

> "No, not at all. In fact, we started sleeping in different rooms as soon as this top search stuff got so intense."
>
> "Ah, that's quite wise! Quite wise! Thanks, that's reassuring."

Brenna was then not so sure that Alvin was really kidding, but she could not figure out how to explain his behavior. At least this is what she remembers, 20 years later.

To relax the nerves of top seekers, an agreement was struck by the spokespersons of the two collaborations, sponsored by Fermilab director John Peoples. Before going public with the announcement of a definitive discovery, each collaboration would grant a grace period of one week to their competitors. This would allow the latter to put together their own case for a share of the merit for the discovery, if their data granted that possibility.

The agreement looked foolish to a few CDF members. Why, they reasoned — that is how science is made: you get there first, you publish. Or, perhaps, they felt strongly about this also because, at least in part, they were certain that the agreement was a concession to DZERO rather than an opportunity for CDF. Others felt it was a good deal: it avoided the scenario where DZERO, upon chancing to get two or three clean dilepton events in a row, decided to go public with a discovery announcement, getting all the press and leaving CDF pants down. Albeit unlikely, that was a frightening thought. And indeed, one week was very little time. It would not be enough time, e.g., to complete a reconstruction of additional raw data queued up to be analyzed, in the hope of finding a few extra top candidates. Usually, raw data would take a few months to be made available to analysts after a full reconstruction and calibration, from the moment it had been collected by the online output streams. On the other hand, one week would be enough to polish off and wrap up a half-baked result.

JETVTX or SECVTX?

A fierce battle was fought on the choice of which secondary vertex-finding algorithm to use for the top search in single-lepton events. Thanks to an idea of Joe Incandela, who was then a co-convener of the b-tagging working group, and the painstaking effort of Weiming Yao, a quite skilled Berkeley researcher, a new algorithm had been developed to reconstruct vertex b-tags in jets, SECVTX. The idea of SECVTX was to "seed" the search for a secondary vertex in the jet using two-track vertices, adding to those a third track when possible. A three-track vertex was then a very clean signature of the decay in flight of a b-hadron; when no three-track vertex was found using tracks passing loose quality criteria, one could still settle on a displaced two-track vertex. That was still a powerful way to identify b-quark jets, provided that the tracks pass more restrictive quality criteria — a higher p_T, a more significant impact parameter, and a removal of decay products of long-lived light hadrons.

The b-tagging algorithms were complex tools, and they relied on a number of user-defined parameters which affected their performance. There was thus the danger that one would be led to choose those settings based on how strong a top signal the algorithms would extract in real data. In order to avoid this illegal "fine-tuning" Joe Incandela and Dave Gerdes, who convened the b-tagging group, demanded that the developers not look at the new data while finalizing their algorithms. A decision on which method offered more sensitivity to the presence of top quarks in single-lepton events had to rely on objective criteria based on background rejection and signal efficiency. Background rejection could be tested in generic jet events: assuming those jets only came from light quarks or gluons, the smallness of the fraction of b-tagged events found in those data indicated how powerful the rejection was. To determine signal efficiency, one could instead study how frequently secondary vertices were reconstructed in b-quark-enriched samples of jets, such as jets containing SLT tags, or checking the result with Monte Carlo simulations. The studies were overall quite complex and the results not always straightforward to interpret. The matter generated lasting disagreements among the analysts. Some of the issues would continue to be ground for hot debates until the end of Run 1, as will be discussed in Chapter 11.

At the end of the day, SECVTX appeared to grant a stronger background rejection than JETVTX for a given signal efficiency. However, its application to the data collected until October 1994 showed slightly poorer results than the older b-tagger: the excess of top-like events that SECVTX found in the first 20 inverse picobarns of data collected since the start of Run 1B was smaller. CDF appeared to have been collecting fewer top events; SECVTX failed to b-tag events as frequently as it was expected to do. This looked like a statistical fluctuation: events appeared to have failed the SECVTX selection cuts by small margins. Yet the issue caused many collaborators to reconsider the matter: maybe sticking with the old tagger was the right idea, as JETVTX gave a larger top signal on the data until then collected. They needed to be reminded that the choice of b-tagger could not be an *a posteriori* one based on their performance on real data, as that would bias the results. The decision eventually taken by the group was to "blind" the data again while the optimization of the SECVTX parameters continued.

On December 7, 1994 the final decision on which b-tagging algorithm to use for the new top search was taken at a crowded Top meeting. Paul Tipton, the soft-spoken leader of the JETVTX team, had pre-instructed his Rochester colleagues to keep quiet and avoid quarrels: "Keep your cool — don't get upset, be perfect gentlemen." But he ended up being the one losing his patience! He suspected that the SECVTX team had illegally been tuning the algorithm such that it would show the best possible performance on the control samples of data used to extract the efficiency estimates. A heated debate ensued. Finally, Lina Galtieri presented the set of optimized selection cuts of SECVTX. She proposed to vote on their adoption as the main b-tagging method and to use them on the data so far collected, such that results could be shown at the forthcoming collaboration meeting in January. The motion was approved by the group: the only vote opposed to it was Tipton's.

In the meantime, the integrated luminosity had more than doubled. At the January collaboration meeting, the results of the counting experiment were disclosed in two back-to-back talks by Joe Incandela and Weiming Yao. On the large dataset collected until then, the counting experiments alone now showed an excess collectively amounting to 4.8 standard deviations, and the reconstructed mass distribution of the b-tagged W+4 jets events was impressive (see Figure 1). The top quark was

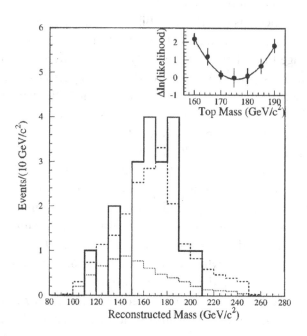

Figure 1.   The reconstructed top-mass distribution in the single-lepton sample collected by CDF at the time of the discovery announcement. The full histogram shows the distribution of the 19 selected data events; the dashed histogram shows the sum of top and background contributions expected for a top mass of 175 GeV; and the dotted histogram shows the expected background contribution alone. The inset shows the likelihood of the data as a function of the top-mass hypothesis. Reprinted with permission from *Phys. Rev. Lett.* 74 (1995) 2626.

nailed down! The writing of a new article titled "Observation of top-pair production at the Tevatron" was immediately started.

Discovery!

At the end of February, the CDF spokespersons informed John Peoples as agreed, and the latter immediately alerted the spokespersons of DZERO: a PRL article claiming discovery of the sixth quark would be sent out by CDF in exactly seven days. The top seekers of the DZERO collaboration undoubtedly had one frantic week. Yet, magically, in the end they managed to produce a paper which itself claimed a five-standard-deviation effect in their data! Finally, the two experiments announced jointly the top-quark discovery in a press release at the Ramsey Auditorium, the conference room in the basement of Wilson Hall.

The spokespersons of CDF were Bill Carithers and Giorgio Bellettini; the latter had recently replaced Mel Shochet. Carithers and Bellettini alternated on stage to explain the CDF analyses, which were now really convincing. Carithers showed the results of the counting experiment. Bellettini then discussed the mass measurement and the kinematic characteristics of top-candidate events. The top mass was measured at 176 GeV, in agreement with the former less precise result; the reconstructed mass distribution was unquestionable by itself. Carithers could also demonstrate that all the characteristics expected from top events correctly checked out.

The hardest part of Carithers's presentation was to explain the curious disagreement between the top-production rate published in the 1994 "evidence" paper and the one that CDF was measuring in the larger data sample: in 1994 CDF had measured a cross-section of 13.9 picobarns, albeit with a large uncertainty, while the new result was 6.8 pb, i.e. half the former one! Carithers showed a plot which would come to be known as the "incredibly shrinking top." It represented graphically a series of corrections to the former result, which all contributed to reduce the estimate of the signal cross section.

Usually, independent systematic uncertainties affecting a measurement are *added in quadrature* to one another to estimate their combined effect. The sum in quadrature consists in adding the squares of the numbers and then taking the square root of the result. If one uncertainty is of

3 units, and a second amounts to 4 units, the sum in quadrature of the two is nine plus sixteen, i.e., 25, whose square root is 5. So, the combined effect of two independent sources of uncertainty on a measurement is not much larger than the largest of the two: being independent, the two effects have an equal chance of mutually dampening one another or producing a coherent, larger shift on the measured value. If, instead, systematic effects are correlated (which means that the value of one influences the value of the other), it is wiser to add the two uncertainties linearly, i.e., 3 + 4 = 7, obtaining a larger total uncertainty. In the case of the top cross section measured by CDF in 1994, all the systematic effects influencing the measurement, while assumed uncorrelated, had conjured together to produce a large positive shift. The sum in quadrature of all considered uncertainties could not account for the resulting bias in the estimate. The "evidence" cross-section measurement thus significantly overestimated the true value.

With the two Fermilab experiments crossing the finish line in perfect synchrony, the top quark was officially discovered. This was a great success for particle physics, although a highly anticipated one. With this result in the bag, CDF and DZERO could now look forward to many other possible discoveries that the new data might bring. For CDF, the difficult path to a consensual view on the way to publish important new results had been beaten, but new challenges were in store, as the future tentative signals of new physics would be much harder to handle than the arguably unsurprising existence of a heavy sixth quark.

Chapter 8

The Impossible Event

Suddenly an uncommon reaction takes place in the detector. It happens on April 28, 1995, in the middle of an otherwise anonymous store. CDF is collecting good data, and the shift crew in the control room takes care of the usual business: keeping an eye on the colorful monitors that plaster the walls, checking trigger rates, logging the warnings issued by the data acquisition system, and answering e-mails.

An Improbable Chain of Events

As a proton and an antiproton run into each other, one red down quark in the proton carries, for an immeasurably small instant, a large fraction of the total energy of its parent. The red down quark gets on a collision course with an antiup quark contained in the antiproton. The antiup quark is also endowed with large energy, and its color is antiblue: in total, the quark–antiquark pair has a net amount of color charge. Yet, before the two bodies get close enough to interact, the antiup quark chances to emit an energetic gluon. The gluon carries away the antiblueness of the antiup quark along with a unit of redness, transmuting the antiquark in an antired one. This allows the now colorless quark–antiquark pair to turn into a W boson, endowed by its parents with a negative unit of electric charge and with energy far exceeding its rest mass. The boson instantly

161

shrugs off some of that extra energy by emitting an energetic photon. Then, it disintegrates, yielding an electron–antineutrino pair. Rather atypically, the electron also immediately emits a second energetic photon.

The proton and the antiproton generating the collision have both been deprived of one of their quarks, so they are now colored, thus unstable. As they leave the interaction point they break apart, creating two streams of low-energy hadrons that fly off along the beam pipe. The energetic gluon emitted by the antiup quark extends the color string that still connects it to the antiproton remnants until the string breaks, yielding two charged pions. One of the two pions receives only a very small share of the energy and ends up spiraling within the beam pipe. The other pion is conversely quite energetic, and it heads straight into the plug calorimeter after leaving a trace of its passage in the SVX.

Once it reaches the calorimeter, the pion withstands a peculiar reaction: it impinges on a lead nucleus where it transfers its up quark to a neutron, receiving a down quark in exchange. This turns the charged pion into a neutral pion, while the neutron becomes a proton. The former lead nucleus, now turned into bismuth, immediately breaks apart into lighter nuclear fragments. The neutral pion only manages to tread five microns or so in the lead slab and then decays into two photons; these in turn produce an electromagnetic cascade which generates light flashes in the scintillator of the calorimeter. The charged pion has performed an illusionist's trick, one dreaded by experimentalists: a reaction called *charge exchange*. What is observable in the detector is a track in the silicon layers, pointing to an electromagnetic energy deposit in the calorimeter. Such a combination is hardly distinguishable from the signal that would be expected from an energetic positron.

Let us now return to the other four energetic particles produced by the W boson: the real electron, the antineutrino, and the two photons. They move out of the interaction point in different directions, heading toward the tracking system. The antineutrino zips unhindered through the detector, leaving no trace of its passage. The electron leaves a stream of ionization in the gas of the central tracking chamber, and once it enters the calorimeter it leaves a well-localized energy deposit. As for the energetic photons, they meet a similar end: they traverse the CTC unseen, but as soon as they enter the calorimeter they initiate additional electromagnetic showers. From the amount of released light in the scintillators at the locations where showers have taken place, experimentalists will be able to estimate the energy of the electron and photons.

After the above confusing chronicle, it is useful to take stock. What remains of the hard collision is the signal of one real electron and two energetic photons, plus a further spurious electron signal originated by the charged pion. In addition to this, the W decay antineutrino has left the detector carrying away a significant amount of momentum unseen, so the momenta of observed particles do not add up to zero in the plane transverse to the beams direction. There is thus a momentum imbalance, a significant amount of missing transverse energy which betrays the antineutrino escape. All in all, what experimentalists have in their hands is a spectacularly improbable event: one with two electrons, two photons, and a significant amount of missing transverse energy. It is going to be dubbed *e-e-γ-γ-met event* by CDF physicists, but a better name for it would be "the impossible event."

Ockham's Razor and Carithers' Jewels

In the 14th century William of Ockham, an English Franciscan friar, philosopher, and theologian, enunciated what he called *lex parsimoniae*, a law of parsimony which he thought was a useful guiding principle in the attempt to understand the world by philosophical means. The principle is often reported in Latin as "*entia non sunt multiplicanda sine necessitate*" (entities should not be multiplied without necessity). It states that the explanation of observed phenomena should not require one to conjecture the existence of unnecessary new entities. We should rather favor an explanation in terms of the fewest possible causes: the most economical explanation, involving fewest assumptions. Ockham's Razor, as this principle was dubbed, was soon recognized as the correct mindset in scientific investigation, a quite effective way to cut down imaginative options and keep on a rational track in the formulation of hypotheses.

Seven centuries after the enunciation of the *lex parsimoniae*, physics rather than philosophy is used in the investigation of the world, and particle physicists are among the most scrupulous followers of William of Ockham. If they observe a peculiar subnuclear reaction, they initially attempt to explain it using known physics rather than the intervention of new, as-yet unknown states of matter or interactions. This is a principled and sound way of doing science.

The chronicle offered at the beginning of this chapter of a chain of standard-physics reactions which could have led to recording the impossible event is just one among several mundane hypotheses of the origin of what was detected. If you polled the members of the CDF collaboration, asking what they thought originated the event, you would get a score of different possible explanations, none of which are too convincing.

To explain the correct way to appraise the rarity of the "impossible event," we may follow the reasoning of Bill Carithers, one of the clearest thinkers in the CDF collaboration. At the end of May 1996 Carithers, then co-spokesperson of CDF, presented an intriguing talk during the "New Physics" session at the topical conference on Hadron Collider physics held in Abano Terme, near the volcanic hills close to Padova, in northern Italy. In his presentation, Carithers stressed the fact that when one collides hadrons, one expects hadrons to come out: by far, the most probable way by means of which a proton interacts with an antiproton is the strong force, and the strong force does not know anything about leptons, since leptons do not carry color charge. Particle physicists are of course interested in strong interactions of very high energy, but they are arguably more intrigued by the more infrequent weak and electromagnetic interactions. The latter may provide a deeper understanding of the subnuclear world and possibly clean signatures of new physics. Collisions yielding energetic electrons, muons, or significant missing transverse energy are therefore of special value, as those are recognizable final states of W or Z decay.

Carithers likened rare particle signals to precious stones in a jewel. While the clean signal of an isolated, energetic electron can be likened to a diamond, a significant amount of missing transverse energy should be compared in value to a ruby, due to the fact that it usually, but not always, betrays the presence of a rare and valuable energetic neutrino. The presence of more stones is bound to raise the jewel's value, except that often these come in pairs: when a jewel has a diamond/electron, it is quite likely to *also* have a ruby/neutrino, because the two particles together are the result of a W boson decay: you never get a single electron alone! Another example would be the decay of a Z boson to a pair of muons: muons are as rare as electrons, so that event should correspond to a jewel with two diamonds; yet, the presence of one diamond in that case demands the presence of a second one.

A further important observation is that when one selectively triggers on special events at a hadron collider, one has to take for granted that the collected events are rare to begin with. You will recall from the discussion in Chapter 2 that the data-recording frequency in CDF cannot exceed few tens per second, while the number of collisions is of the order of a million per second. Physicists naturally arrange to record events which contain "precious stones": that is the very purpose of the complex triggering system. Hence in appraising our event-jewels, we should take for granted the presence of not one but *two* precious stones. Any *additional* stone, however, would significantly raise the jewel's value.

The above reasoning led Carithers to propose a qualitative classification scheme that allowed one to get a feeling of the relative value of the signatures of rare particle reactions. He proceeded to show to his audience some of the most intriguing, hard-to-explain events that CDF had collected during Run 1. Thanks to his appraisal scheme, to each event he could attribute a relative value, due to the rarity of the precious objects it featured beyond the first two.

Following Carithers, we may take for granted the two rarest stones in the e-e-γ-γ-met event, the jewel we are appraising: the two electrons. We are left with three additional quite valuable objects: two energetic photons and large missing transverse energy. Already two of them together would make the event a jewel of the rarest kind, but three make it an unheard-of occurrence. Indeed, in his talk, the CDF spokesperson emphasized the surprise value of the e-e-γ-γ-met event and its potential importance in opening the way to new physics discoveries. If standard model processes cannot produce such a signature, one is justified to look for exotic scenarios wherein the simultaneous production of those five objects may find a reasonable explanation.

To have a hunch at possible standard model explanations of our observation, we take known physical processes that may produce some of the observed particles, and then consider what other accidental sources may have produced the remaining ones. $W \rightarrow e\nu$ and $Z \rightarrow ee$

decays are the starting point: they both provide an economical explanation of two of the jewels observed in the anomalous event. *Diboson production*, which is the coherent production of a pair of vector bosons (W, or Z, or photons), goes one step further at the cost of a significant increase in the rarity of the event. Each of the decays W$\gamma \rightarrow$ e$\nu\gamma$, WW$\rightarrow$ eνeν, and Z$\gamma \rightarrow$ eeγ explains three of the five objects (the two neutrinos of the WW final state add up to produce large missing transverse energy in most kinematic configurations). In the first case, we are left with the task of explaining a second electron and a second photon signal, in the second we need to explain the two photons, and in the third the missing energy and the remaining photon. For example, we can explain everything with a four-boson final state such as WW$\gamma\gamma$, but that is a bit like throwing the towel, as the standard model predicts that an event of that kind should appear in Tevatron collisions less than once in a million years.

We thus look into other mundane explanations, keeping diboson production as a baseline. At least one of the two additional objects might be due to some coincident spurious process. Cosmic rays depositing energy in the calorimeter simultaneously with the proton–antiproton collision are a possibility, although a very unlikely one. Another one is instrumental noise, like a photomultiplier tube in the central electromagnetic calorimeter "firing up" without cause: a spurious photomultiplier pulse would be interpreted as a large localized energy deposit unmatched to charged particles pointing at it, making a clean photon candidate. A third possibility is that a second proton–antiproton collision took place during the same bunch crossing, yielding an additional rare electroweak process. This could, for instance, be the production of a pair of photons, or a photon recoiling against a jet which generates the spurious signal of an electron by charge exchange, similar to what is hypothesized in the discussion at the start of this chapter.

Any way one turns it, the problem remains. Known physics appears unable to offer a credible explanation to the impossible event. Having exhausted possibilities predicated on current physics, we can reasonably entertain speculations of new entities. Ockham's Razor can be returned to the bathroom cabinet, as we need to explore possible explanations that require new physics.

You've Got Mail

Given the dozens of different output data streams into which the CDF data acquisition system was divided, and the large number of events written into each dataset during Run 1, one might imagine that the unearthing of the impossible event took time and some careful sifting through a large number of less interesting proton–antiproton collisions. It would make indeed a nice story if it were so.

The protagonist could be Jane, a sleep-deprived graduate student working on an innocuous measurement of the Z boson cross section. Before leaving for the evening, Jane's advisor has suggested her, among a number of other things, to check the energy balance of her candidate events. Accordingly, she is now producing a histogram of the missing transverse energy for events containing a pair of electrons, trying to get it done before rushing out of the lab to meet her friend at the movie theater. The histogram she added to the analysis job running on the dielectron dataset extends from 0 up to 40 GeV: the upper boundary is high enough that the histogram should contain all her events. However, upon opening the job output and checking the graph, she notices that the "stats" box printed by the program at the top right declares the presence of one over-flow entry. One event in her sample has missing transverse energy exceeding the 40-GeV bound.

Jane scratches her head; she remembers a time when her advisor yelled at a colleague that overflows must not be ignored, ever! She decides to regenerate the histogram using a wider range. It takes 10 seconds to change the upper boundary of the histogram in her code, 20 minutes to compile the code and produce a new executable job, and one full hour to re-run it on the data. That means canceling the movie appointment, but she is accustomed to not having a life. She has sorely come to accept that a normal life is a commodity which graduate students have no access to.

As the job finishes running, Jane opens the new output file, and this time she spots the odd outlier entry with 53 GeV of missing transverse energy. What the heck is that outlier? None of the other 10,000 dielectron events in her sample have a missing transverse energy above 30 GeV or so. She is now genuinely intrigued, and her regret for the lost movie is behind her back.

She wants to know what this event is like. A third spinning of the dataset finally allows her to select the event from the database, so that she can now examine its characteristics in detail with the event display program.

Incidentally, the event display in Run 1 was called DF, as in "Damn Fast." It was the creation of Bill Foster, who in Run 0 had eventually lost patience with the commercial product originally adopted by CDF, "DI3000." DI3000 was a cumbersome software which produced a three-dimensional visualization of the detector, but took hours to run. Bill had decided to trash it and write a leaner and meaner display program from scratch, DF. Everybody had since then been enjoying it a lot.

Jane opens DF and launches the calorimeter display. Bang! What are those two extra electromagnetic signals in the central calorimeter? A check at the CTC view reveals no tracks pointing toward them. Not only does the event have an abnormally large missing E_T, but it also has two extra photons! Can it be real? This must be new physics! The graduate student picks up her office phone; her hand hovers on the keys as she glances at the wall clock. Hesitant, she bites her lip: it has become quite late. But then out of an impulse she dials the number of her advisor. The sleepy voice answering her call explains with a gruff voice that no, her advisor is not home yet — he has gone to the movies with friends....

Alas, no. The way the impossible event was discovered was not that picturesque. Shortly before Run 1A David Saltzberg, then a Ph.D. student from the University of Chicago, had been prompted by his advisor to write an event-scanning program which would monitor the collection of selected signatures. The code was slavishly called *PhysMon*. Every time CDF started data taking during a physics store, PhysMon was brought up to run together with all the other software modules that made up the Level-3 trigger. It had been originally conceived as a monitoring tool to signal the presence of anomalous events. These could be due to the malfunctioning of detector components or a failed reconstruction of event properties. However, it soon doubled up as a precious instrument that could alert the researchers of the collection of odd and rare physics processes.

The program ran on the *Express stream*, a dataset temporarily stored on a pool of storage disks. The Express stream collected some 10% of the total output data, *la crème de la crème*: events with high-E_T electrons or muons, or large missing transverse energy, or high-E_T jets. The data at that stage had not yet been subjected to an offline reconstruction and calibration: only the output of a speed-optimized reconstruction performed by the computing farm of the Level-3 trigger was available.

Using the raw information of the Level-3 reconstruction, PhysMon looked for events that contained striking signatures: precious jewels, in Carithers' jargon. To define what to call a jewel, one had to adjust the thresholds in transverse energy above which the different gems could be considered: electrons, muons, missing energy, very high-energy jets. That made sense, since it was the large transverse energy of the particles what made them rare and valuable. As for photon candidates, they were not in the VIP list: back then, photons were not seen as a possible worthy signature of new physics or other fancy processes, mainly because of the large background that photon candidates suffered from hadronic jets. However, the simultaneous presence of a photon candidate and missing transverse energy was useful to monitor because it signaled an occasional malfunction of the CTC. Those were most likely W decays to an electron–neutrino pair; while the neutrino produced genuine missing E_T, the electron signal in the calorimeter was interpreted as a photon if there was no reconstructed track pointing at it. This mostly happened when an interaction from a *satellite bunch* of protons produced a $W \rightarrow e\nu$ decay. Satellite bunches were packets of protons that interacted in the center of CDF very slightly out of time with the electronic readout of the detector; the time lag (usually of the order of 20 nanoseconds) threw the track reconstruction out of whack. Saltzberg used to forward those events to Peter Berge for a cross-check, and Peter invariably answered by telling David he had won the "satellite bunch award of the day."

If an event passed some predefined "oddity criteria," PhysMon generated an automated email message to a short mailing list; the event file was attached to the message. The list members were accustomed to receiving a trickle of such messages during data taking, maybe two or three per day. What they were not used to was receiving four consecutive emails with the same subject, "Run 68739, event 257646." The impossible event was so strange that it passed four distinct sets of PhysMon criteria, and three of them regardless of the two extra photons! It generated an email because it had two electrons with large total invariant mass, but it was also singled out by having two electrons and significant missing transverse energy; further, the combined "transverse mass" of one of the electrons and the missing energy was very large, and this was also a criterion which made the event worth flagging. And finally, the photon plus missing E_T flagged a possible track reconstruction failure.

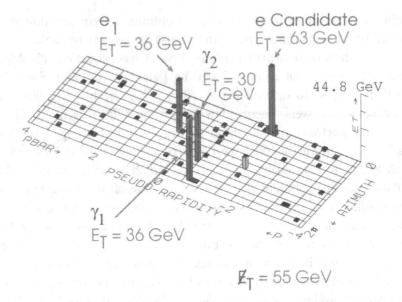

Figure 1. DF display of the transverse energy deposits left in the calorimeter cells of CDF by the e-e-γ-γ-met event. The plane is described by azimuth (an angle in the plane orthogonal to the beams) and *pseudo-rapidity* (a function of the angle in the longitudinal direction). The height of the bars is proportional to transverse energy; the highest ones are the photon (γ_1, γ_2) and electron candidates.

As soon as the event hits the mailbox of the members of the Chicago group, they understood its surprising nature. They printed out the colored three-dimensional "lego" plot showing energy deposits in the calorimeter (see Figure 1) and the transverse view of observed tracks. They stuck the two graphs on the wall of the corridor where other event displays had already found home, and started to think at the possible explanations. Multiboson production came to mind, as well as some erratic failure of the tracking system to correctly match tracks to the electromagnetic deposits in the calorimeter. Yet, the expected rate of such processes appeared too small to constitute a valid explanation. That event was inexplicable. It might indeed be a first hint of new physics, maybe supersymmetry!

While it was getting the CDF members excited, the impossible event also quickly made it outside of CDF. The first public use of the event was made, in fact, only two weeks after it got collected, during a talk that Seongwan Park, a post-doctoral scientist from the Fermilab group in CDF, gave at the 10th topical workshop on Hadron Collider Physics in May 1995. Back in those days, it was considered acceptable by CDF to show event displays outside the collaboration without first subjecting them to a formal blessing procedure, so Park could show the event as a *divertissement* at the end of his presentation, which generically focused on searches for new exotic particles. At the conference, nobody assigned the oddity much relevance. Apparently, theorists overlooked it or did not take it as an inspiration to cook up fancy models which could account for the phenomenon. It would take one more year for that to happen.

The Other Chicago Zoo

It was the impossible event what really caused the group of physicists from the University of Chicago to gradually phase out some of their ongoing analysis efforts to concentrate on a wide-range, coordinated search for exotic signatures including photons and leptons. Their interest in particle zoology had been there even before the start of Run 1, when PhysMon begun sending its emails. The walls in the corridor of the physics department where the CDF members had their office were plastered with colorful event displays of the most unusual collision events: these included events with abnormally high-energy electrons or muons, multijet events with large missing transverse energy, or multilepton events of all kinds.

Over time, with the increased integrated luminosity, some events initially considered abnormal or exciting had begun to look stale, and had been replaced with newer, rarer ones. That paper menagerie had grown larger as Run 1 progressed. The displays had an inspirational function: the researchers would stare at them, trying to capture their salient features and creating some sort of data bank in their minds. That way, they would learn to distinguish similar configurations in new events they chanced to see.

In the late spring of 1995, a "Zoo-fest" event was held at Fermilab. It was a half-day meeting centered on the oddities that CDF had collected during Run 1, a sort of "show-and-tell." The idea was to go over those odd events, estimate their rarity, and do some brainstorming over their characteristics and the possible indications they were giving on the potential presence of new physics in CDF data. A few Fermilab theorists were invited, too. At the meeting, Tom Lecompte showed the e-e-γ-γ-met event, along with some back-of-the-envelope estimates of the rates of standard model processes that could yield a similar signature; those were all dismayingly small. Dave Toback, who had recently joined the Chicago group as a Ph.D. student, was strongly impressed by the event. Toback's initial studies had concentrated on W plus jets events, but he was about to drop that topic and steer toward a way more exciting subject for his thesis.

On the following week, Toback attended a 4 pm seminar at the Enrico Fermi Institute in Chicago. The speaker was Francis Halzen, a University of Wisconsin professor and a veteran of particle physics. Halzen was talking that afternoon about a new detector for ultra-energetic cosmic rays in construction at the South Pole. Although the seminar was quite interesting, Dave could not focus on it: his mind kept going back to that e-e-γ-γ-met event he had heard about a few days before. Was it new physics? How to investigate it? Finally, it dawned on him: of course, cousins! One could easily set up a search for cousins of the e-e-γ-γ-met event, like the search for events similar to the "monster Z" event that his group had done just one year before! Dave spent the rest of the seminar anxiously waiting for its end: he needed to get a hold of his advisor that afternoon.

The "Monster Z" event was another early member of the other Chicago Zoo. Collected during Run 1A, the quite peculiar event had triggered a short study at the end of 1993. It featured an energetic electron–positron pair fully compatible with a Z boson decay, recoiling against two jets of extremely high energy. Although rare, the production of a Z in association with hard QCD radiation was a legitimate explanation of the observed characteristics of the "monster Z." Yet, one could caress the hypothesis that a new exotic particle of very large mass, say 1 TeV, had decayed to a Z boson plus a second new particle, and the latter had produced the two very energetic jets. What the Chicago researchers had done back then was to search for "cousins" of that event, ones which in the place of the Z → ee decay featured a quark–antiquark decay of the Z; the latter is 20 times more frequent than the former. Had CDF collected a score of cousins in the 4-jet final state? No: the Chicago group had found no similar events which could indicate a common exotic source. That anomalous event had been archived as unexplained, fancy, but fruitless: a legitimate member of the Zoo.

Brainstorming over the impossible event, Toback reasoned that although PhysMon did report events with photon candidates, nobody used to care about them, due to the significant background from cosmic rays and QCD jets affecting their signature. Yet, the photons were exactly what was most striking about the impossible event. The two electrons of high invariant mass and the missing E_T were a golden signature of electroweak processes, but it was the addition of the photons that placed this event in a different league. One could consider processes that made photons appear together with W or Z bosons, or with other physics signals yielding two electrons and missing E_T. The electrons were only one of the possible signatures of vector boson decay that CDF could spot. There might be multiple jets, or an electron and a muon, or other mixed signatures that maybe would not appear as striking by themselves. Again, cousins of the e-e-γ-γ-met event.

The Chicago group had been proactive with the early catching of interesting signatures: the spin cycle, the Express stream, PhysMon. It had been a long-standing, coherent effort to get early warnings of potential new physics in the data. What Dave was focusing on was one small but significant weak spot of the whole system: photons were not considered a clean signature of potential new physics; they were "semi-precious stones" in the jargon of Carithers' analogy. And photons were special if one tried to fit them in Carithers' appraisal scheme. Two gems per jewel could be given for granted in the case of electrons, muons, missing energy, or even with energetic hadronic jets, because those "gems" were produced in pairs in the decay of W and Z bosons, or, in the case of jets, in the energetic scattering of two quarks or gluons. With photons, this was not the case. If genuine, one single photon was already an interesting signature; and together with electrons and muons, it flagged interesting multiboson production, a subject studied little at the Tevatron until then.

The conclusion of Dave's brainstorming was clear. One had to attack the problem by searching similarly striking events: photon-pair candidates appearing together with electrons, muons, or other precious gems. Until then, Dave had been working on a study of W + jets events, but the impossible event was much more exciting. The W + jets sample would have to wait; he would come back to it a few years later. From now on, Dave's work was going to be the deep investigation of all known and unknown physics that could produce final states containing photon pairs and anything else worth reporting. This would eventually become the subject of his Ph.D. thesis. The search for cousins of the impossible event was on.

In Search of Cousins

The matter demanded scientific rigor. All possible explanations of the e-e-γ-γ-met event had to be scrutinized. Quantitative estimates of the predicted rates of similar events in the data collected until then had to be tabulated. Only at that point could one start entertaining the hypothesis of sources that laid beyond the standard model. The two researchers who took on this task were Ray Culbertson and Dave Toback. Ray had recently joined the Chicago group as a post-doc. He had previously worked on a fixed-target experiment at Fermilab and he had seen the Chicago group as the right team to join to get involved in searches for new physics. However, until then rather than the search for new physics Ray's work in CDF had been the precise measurement of the *Drell-Yan* cross-section. The Drell-Yan process is the one by which a quark in a hadron, colliding with its antiparticle in another hadron, annihilates into a photon or a Z boson. This was clearly a process of extreme interest, due to the rather clean and theoretically calculable physics it gave rise to, but it was not a process likely to yield any big surprise or a new discovery. So, the offer of making a U-turn to concentrate on the chase for extremely rare events was an attractive one to him.

In the matter of a few months, the Chicago researchers spun all the datasets that could contain *diphoton* events (i.e., ones with two photons), analyzed the selected data, and prepared a draft of a CDF note. The document, while far from addressing all the nagging questions that could be asked about the odd event, did contain a quantitative assessment of one specific scenario. The authors reasoned as follows: if the event has the appearance of being due to WWγγ production, then let us assume that it indeed contains those particles in the final state. If that is the case, then whatever is the physical process that originates such an odd combination of four bosons, we should expect to see in CDF data additional WWγ γ events yielding different signatures, according to the other decay modes of the two W.

As discussed in Chapter 4, out of nine W boson decays one expects on average one electron–neutrino pair, one muon–neutrino pair, one tau–neutrino pair, and a total of six jet pairs. Now if the e-e-γ-γ-met event is caused by the decay to electron–neutrino pairs of *two* W bosons, this is just one of 81 possible decay modes of WW pairs. So, one expects that the same data sample where the impossible event was found contain on average one μ-μ-γ-γ-met event, as well as two e-μ-γ-γ-met events (since there are two ways to produce the mixed electron–muon signature: the first W can give an electron and the second a muon, or vice versa). But crucially one also expects to see as many as 36 γ-γ-jet-jet-jet-jet events. A search for those events, however, returned none. Diphotons plus missing energy plus only one electron, or one muon? None. Diphotons with large missing energy? Zilch. Diphotons plus four jets? Zero. The absence of cousins of the e-e-γ-γ-met event was a confirmation of its rarity. And as the draft CDF note describing that result looked polished enough to be ready for collaboration-wide distribution, CDF received a letter which would give further motivations to the Chicago group efforts.

Sleeping on the Job?

1995 had been an eventful year for CDF: the top discovery had cast the experiment under the spotlights worldwide, the Tevatron had delivered almost 100 inverse picobarns of collisions, and experimentalists had been extremely busy on a number of fronts. The high instantaneous luminosity that the accelerator had been providing toward the end of Run 1B had put pressure to all the scientists who operated the detector. Under high-luminosity running conditions, five lost minutes of data taking meant losing on average one clean Z decay to lepton pairs; and three hours of down time were enough to lose a top-pair event. In order to collect all the data they could, shift crews alternated day and night in the control room, keeping all systems up and running normally, and promptly reacting to their occasional hiccups. Trigger settings were constantly revised in order to adjust them to the changing running conditions and to keep the dead time below the psychological threshold of 20%. Subsystem experts carried pagers 24/7 and spent sleepless nights attending to their detectors, ensuring a faultless operation. The members of the Offline group were also severely burdened, since the large flux of data called for an increased effort to produce calibration constants, reconstruct the events, and make "production datasets" quickly available to analysts.

And then there was the production of scientific results. A team of 4–5 researchers could work full-time for one or two years to pull off a measurement worth being published. Once formally blessed, the result would be reviewed by three "godparents" assigned by the spokespersons. These internal reviewers would request further verifications, changes, additions, checks. Then the drafting of a physics article would follow along with a phase of "collaboration wide review" when every Joe or Jane would have the chance to raise questions. Everything had to be answered in good order, supporting material had to be provided to anybody who asked for it, and innumerable tweaks had to be made to the text before submitting it to a scientific journal. All in all, the completion of a typical CDF publication took a dozen man-years of work! In 1996 alone, 17 such publications were produced.

Despite all the above activities, CDF must have been giving a different impression to the outside world. Particle theorists, in particular, while quite interested in the new observation of the top quark and in the measurement of its properties, were more intrigued by the possibility that Tevatron data could finally break the standard model. Most phenomenologists were especially interested in the many signatures of supersymmetric particles that CDF and DZERO could detect: supersymmetry was already all the rage in those years. They thus looked at the Tevatron with a mix of apprehension and hope. The Fermilab experiments were producing several searches of supersymmetric particles such as charginos, neutralinos, gluinos, and squarks, yet those analyses appeared insufficient: supersymmetry could be searched for in literally dozens of independent ways. It had to be there and the Tevatron was not doing enough to unearth it!

The matter came up one summer evening of 1995 in a conversation at Opal's restaurant in Santa Barbara, where Pierre Ramond, Joe Lykken, Gordon Kane, Mary Gaillard, and Paul Langacker were having dinner. The five theorists were participating to an ITP workshop at the University of Santa Barbara titled "Unification: from the Weak Scale to the Planck Scale." The workshop had the goal of gathering together string theorists and model builders to create the grounds for a coherent effort in the production of new ideas in the field. Unfortunately, the plan failed, since all stringers were huddling off by themselves the whole time, talking about the then newly discovered "string dualities," a recent exciting development in string theory. One thing united all participants to the workshop, though: they were completely sold on the idea that supersymmetry was the correct way to extend the standard model. In spite of the dearth of experimental hints in that direction, the theoretical benefits of SUSY were hard to ignore. It looked like no other theory could match the elegance and depth of the enlarged symmetry of a supersymmetric world. SUSY was also very welcome by string theorists, as the correctness of string theory could be argued to imply the existence of a supersymmetric copy of standard model particles.

In the face of all those theoretical motivations, it looked as if too little emphasis was given to SUSY by experimental searches at the Tevatron. To make the matter more urgent to US physicists, Pierre Ramond reported that he had just flown in from Europe where he had witnessed the strong excitement of CERN physicists for the ongoing upgrade of the LEP collider to LEP II. The increased center-of-mass energy of the electron–positron collisions about to start being collected by the four LEP experiments promised a lot. CERN would soon give itself a fighting chance of putting its European hands around the coveted trophy of new physics beyond the standard model, by being the laboratory where new supersymmetric particles were discovered!

While LEP II was anticipating those promising new datasets, the Tevatron had already a much broader territory to investigate at its arms' reach: CDF and DZERO were sitting on a huge pile of data where hidden treasures might be hiding. The thought that the two Fermilab experiments were not investing enough efforts in SUSY searches was disturbing. The five colleagues that evening found out they fully agreed that it would be a shame for particle physics in the USA if the chance of the discovery of the century were stripped from the leading center of particle physics in America and offered on a silver tray to CERN.

Something had to be done. Ramond put forth the idea of writing a letter to the Tevatron experiments; this found everybody enthusiastic. The five theorists spent their evening laying down the scheme of the text and a list of prestigious names who could be asked to add their signatures below it: Jonathan Bagger, Vernon Barger, Howard Georgi, David Gross, Lawrence Hall, Frank Paige, Roberto Peccei, Joseph Polchinski, Steven Weinberg, Frank Wilczek, Edward Witten, and Bruno Zumino. Back from dinner, Ramond and Lykken started to draft the letter. After a few revisions, the letter was sent to the 12 potential subscribers identified at the dinner.

All of them agreed to sign it, except Howard Georgi. And so it was that in the fall of 1995 the letter reached the hands of Fermilab director John Peoples, as well as those of the experiments' spokespersons Paul Grannis and Hugh Montgomery (DZERO) and Bill Carithers and Giorgio Bellettini (CDF). The text, dated November 15, was a statement of the belief of the 16 subscribers:

> "As the union of special relativity with quantum field theory sixty years ago predicted the existence of antimatter, supersymmetry [...] predicts a new form of matter, the superpartners of the elementary particles. It is our belief that the measured values of many of the standard model parameters point to their existence [...]:
>
> • The large value of the top quark mass [...] yields a natural explanation of electroweak symmetry breaking [...]
> • The measured value of the weak mixing angle is linked, through supersymmetric unification, to the existence of superpartners with masses of hundred(s) of GeVs
> • The absence of corrections from new physics in the precision measurements of the standard model is naturally explained by the decoupling of the superpartners from its radiative structure."

Theorists thus urged experiments to invest more efforts in the hunt for SUSY:

> "We, the undersigned, believe that Fermilab has unique detection possibilities for supersymmetry, and urge you to direct your laboratory's efforts in that direction, and ask the leaders of the collider detector collaborations to intensify their search for massive superpartners."

By reading the letter two decades after it was written, one cannot help grinning at the questionable encroachment. Without being asked to do so, 16 influential particle theorists attempted to steer the physics program of the Fermilab experiments toward a broader plan of investigations of supersymmetric signatures. Those studies of course were a good thing in principle. But the resources needed would have to be drawn away from other interesting measurements and searches. With hindsight, the three "experimental hints" that the theorists mentioned in their letter do not look that compelling. Yet, the call to arms had significant consequences for the research program of CDF and DZERO.

Merry Christmas, Gordy!

The letter of the 16 theorists indicated supersymmetry as a priority, so it made sense for the Chicago researchers to add to the note they had been putting together a section discussing possible supersymmetric explanations of the odd event. What was needed was a credible SUSY scenario and a cross-section estimate, or even better, a full theoretical model at least loosely compatible with the observation.

But what could that scenario be? The question, presented to several theorists, returned no hint. Everyone generally pointed to large missing transverse energy as a definite indication that the event could be due to SUSY reactions, but that was a quite well-known and generic element, and one offering very little inferential power. In most of its instantiations, SUSY did predict the existence of a neutralino, a particle which would create missing transverse energy upon being emitted in a proton–antiproton collision. As mentioned in Chapter 1, the neutralino could be a candidate for dark matter in the universe. Instead, nobody could suggest a meaningful model fitting to the strange signature that CDF had observed. But the Chicago members were undeterred. Their reasoning was very simple: photons play a very important role in our world, how could those particles not play a similarly important role in the supersymmetric half of our universe?

Finally, the question was turned to the one and true SUSY guy, Gordon Kane. Kane is one of the most active theorists in the area of supersymmetry, as well as one of its most fervent advocates. Enquired by telephone on possible supersymmetric interpretations of the impossible event on the morning of Christmas, he gave the answer.

"Well, I'll be damned! Of course! That's a picture perfect signature of SUSY! And I've predicted it! I wrote a paper with Howard and Mariano ten years ago precisely on this signature. And Howard and I even wrote a *Scientific American* article on this!"

Howard was Howard Haber, and Mariano was Mariano Quiros. They were two theorists whose research interests centered on supersymmetry, as did Gordon's; they had collaborated in the investigation of some peculiar models in the past.

"Yes, we wrote a *Physics Letter* article in 1985," he explained. "You produce selectrons and they decay to electrons and photinos, and each of the photinos yields a photon and a neutralino. And the *Sci-Am* article ... I believe it's in the June 1986 issue."

The CDF members were asking for a possible estimate of what cross-section those events could have in Tevatron collisions, so Kane looked for a clean sheet of paper and a pencil. Then he started to write down the event characteristics, like a waiter taking a complicated order. The production rate for selectrons depended on the mass one assumed for them, and it was also connected to the probability that photinos would decay to photons and neutralinos. If one wanted the latter to be large enough to make similar events observable, this pointed to some very specific corner of the parameter space, where the whole decay chain could be reasonably likely.

As he hung up, the SUSY guy looked at his wife Lois with an apologetic expression on his face. He would have to spend the holidays working at that thing. Ten years before, he had put in print his hypothesis of a particular signature, one that could happen only if supersymmetry got realized in a special way. Now that an experiment was seeing something that could verify his claims, he had no choice: a meaningful cross-section estimate for the impossible event was urgently needed. Eventually, rather than just affecting his holidays, the construction of models where the impossible event could fit in would keep Kane busy for much of the following year.

Theoretical Ideas and Experimental Tests

Shortly after Christmas, the Chicago group published in the CDF notes archive their search for cousins, which now included a short section mentioning Kane's model and providing the crucial references. In the meantime, Kane's calculations yielded an estimate for the rate of selectron pair production. This turned out to be only a factor of two below what the observing of one e-e-γ-γ-met event in Run 1 data implied. The result thus appeared a nice confirmation that SUSY could explain the impossible event.

Other phenomenological considerations were also suggestive. The absence of μ-μ-γ-γ-met events in CDF data was not a setback for the selectron model, since one just had to assume that smuons were heavier than selectrons in order to explain the lack of those particular cousins. It was also encouraging that the model predicted the absence of e-μ-γ-γ-met events. The lack of those combinations could be read as due to the fact that superparticles were produced in pairs: one could not produce a selectron and a smuon together as this would violate the conservation of lepton flavor. Background processes yielding two W bosons, on the contrary, would produce twice as many eμ combinations as ee or μμ ones — but none had been seen. For SUSY-believers, there was ground to become excited!

Besides a deep study of the physics of diphoton production, there were a number of other ways to address the questions that the extremely rare event was raising. The group worked with the goal of understanding whether the impossible event was the tip of an iceberg or an unrepeatable oddity. Dave Toback concentrated his studies on the signature of two photons produced simultaneously with anything rare enough to be worth mentioning: electrons, muons, missing energy, high-E_T jets, or a free spot in the trailer parking area. Jeff Berryhill, a graduate student who had recently joined the group, later took over a part of Dave's workload, to study signatures which included one photon and one lepton. This would target with higher sensitivity the signature that Gordon Kane had hypothesized.

Ray Culbertson took a slightly different course. He focused on events containing one photon and missing E_T in association with a b-tagged jet. The idea of this search was founded on the special role that theorists attached to the third generation of fermions, to which b-quarks belonged. Particles foreseen by new physics models could be hypothesized to have privileged couplings to the third generation of matter. Also, most realizations of supersymmetry foresaw that squarks displayed an "inverted hierarchy," whereby third-generation squarks would turn out to be the ones of smallest mass and thus the easiest to produce. Their decay was expected to yield the corresponding standard model quarks. In Carithers' appraisal scheme, b-quarks could thus be likened to topaz or aquamarine stones. Instead, the other third-generation quark, the top, while much more precious, would be too heavy to produce in the decay of SUSY objects: it was hard to imagine that exotic processes involving top quarks and photons would yield significant numbers of events.

As the above searches progressed, CDF speakers giving talks at international conferences begun to experience pressure from theorists and experimentalist colleagues of other collaborations. Presenting analyses even remotely connected with the exotic signature of the impossible event usually resulted in a lively question time. The speaker would receive detailed questions on the characteristics of that oddity or on the presence of similar events in the Run 1 dataset. DZERO speakers were often asked whether they too had observed similar events in their data sample, but they had not. The *Cornell Arxiv*, the electronic repository of "preprints" of scientific articles, soon started to receive a significant flux of papers offering explanations of varying degree of improbability for the CDF observation. The event had managed to capture the interest of theoretical physicists more than could the top-quark discovery.

Among the serious models that survived the initial sorting, two stood above all others as ones worth investigating. One was the already mentioned supersymmetric signature of Kane and Haber, which involved pair production of selectrons, followed by the decay of each selectron to electron plus photino, and the subsequent decay of the photino to photon plus neutralino. The other was a "gravity-mediated supersymmetry breaking model" developed in the winter of 1996 by Sandro Ambrosanio and Gordon Kane. In that model the *gravitino*, the superpartner of the *graviton* (the particle supposed to mediate gravitational interactions), was imagined to be the lightest supersymmetric particle. This had phenomenological consequences which included a set of signatures of supersymmetry quite different from the canonical ones. In those models, photons could result from the decay of neutralinos to gravitinos. Theorists soon implemented the relevant processes in Monte Carlo simulations; the latter were used by the Chicago group to search for the resulting signatures.

On the experimental side, the "CDF data" being compared with the models was no longer just the impossible event. In 1996, a search in the full Run 1 data set revealed the existence of additional events that, although not strictly "cousins" of the e-e-γ-γ-met event in the sense intended by Toback one year before, could conceivably fit in some common picture. The Chicago group discussed in CDF note 3571 the e-e-γ-γ-met event along with three others. The first was a "μ-μ-γ-γ-met" event which, although apparently similar to the impossible event, had different kinematic features, as well as a few serious flaws in its "precious stones." The event contained three jets in addition to muons and photons. The missing transverse energy was less significant than that of the original e-e-γ-γ-met event, and the presence of jets implied that a mis-measurement of hadronic energy in the calorimeter could not be ruled out as a source of missing E_T. Also, one of the two muons was on shaky ground since it failed one of the selection criteria that defined good muon candidates. The second event was an electron–photon event which included significant missing transverse energy and three hadronic jets, one of which was "fat" enough to look a bit odd; it was a striking event, but way less so than the first one. The third one was different from all others: it was an event with three leptons, and it was included in the note for its intrinsic rarity rather than its similarities with the others. One of its two electrons, like the muon of the second event, was also not convincing as it failed one of the requirements usually imposed to leptons in standard searches.

All in all, the four events were arguably among the strangest of the whole Run 1. In their conclusions, the authors stressed that at least the first three could fit in a common new physics description. They were, however, careful to explain that one could not rule out mundane explanations. While doing that, they added a mention in the references to the fact that their Argonne colleague Larry Nodulman had famously offered 500 dollars to anybody who would come up with a plausible explanation of the lot. Larry's money was never claimed.

The Other Anomaly: Ray's Dijet Bump

In early 1996, Ray Culbertson found himself fighting on various fronts at the same time. While following Dave's work on diphotons and keeping an eye on Jeff's photon plus electron analysis, he was also working on the sample of single photon events which contained missing transverse energy and b-tagged jets. That study was interesting because of the already mentioned theoretical idea that third-generation quarks could be a door to new physics. Additionally, during the winter of 1996 CDF had been dealing with an anomalous peak at 110 GeV that Paolo Giromini and collaborators had discovered in the mass distribution of b-tagged jet pairs. Those were events selected to contain a leptonic-decaying W signal (the story of that anomaly is told in Chapter 11). One could imagine that a similar effect, if real, could also appear in events with a photon in place of the W, as the W boson and the photon belong the same family of electroweak force carriers.

Ray defined a sensible set of selection criteria and proceeded to work out a comparison of the selected data to the number of events predicted by standard model sources, in search of an excess due to exotic processes. His dataset included all events containing a photon candidate and a significant amount of missing transverse energy. Then he focused on the subset of events containing hadronic jets. Using the Method-one background calculation (see Chapter 6), he estimated how many events in his data should contain a SECVTX b-tag; this yielded a prediction which undershot the observed number. In other words, there was an excess of b-tagged events. One could thus entertain the hypothesis that an extra source of bottom quarks contaminated the sample. Ray, therefore, turned his attention to the kinematic characteristics of the b-tagged events he had collected. For each studied variable, he could compare the histogram of the selected events with the Method-one prediction: this made it possible to distinguish whether the excess of events was coming from a specific kinematic region, or whether instead it was broadly distributed and featureless. The former situation would suggest that the data contained some new signal; the latter instead would point to a simple shortcoming of the background prediction.

A no-brainer distribution to produce using the subset of events containing two jets was the *dijet* mass distribution: a histogram of the combined mass of the jet pairs. If the two jets originated from the decay of an exotic new state produced in association with the photon, a bump would appear in the dijet mass histogram. Ray's mild excess, however, did not pile up at a particular mass value. But then one day, talking to colleagues over coffee, he received a precious hint. The topic of the discussion was the Higgs search that Giromini and the Frascati group were carrying on, and Paolo Giromini's approach to the reconstruction of his tentative resonance. According to one of his colleagues, the most principled thing to do in a search for dijet resonances had been shown years before by the UA1 experiment. A precise measurement of the mass of an object decaying to two hadronic jets required that the latter be reconstructed with a wider *jet clustering* radius than the one used in most CDF analyses after the top quark discovery. Ray was using the standard radius of 0.4 units, while his colleague was suggesting a radius of 1.0, two and a half times wider!

Here I need to explain what jet clustering is. The energy of a hadronic jet is measured as the particles that make it up interact with the dense matter of the calorimeter. To reconstruct the energy and direction of the parton that originated the jet, experimentalists cluster together the localized energy deposits in the calorimeter within circles of fixed radius. The algorithm moves these circles around in the idealized space formed by the calorimeter surface, in search for the configuration which maximizes the sum of energy deposits collected within their bounds. A narrow circle is usually capable of better measuring the energy and direction of the originating parton if this is very energetic and if there are many jets to reconstruct. A wider circle can instead include the trajectory of particles emitted at large angle from the initial parton direction and might be preferable in less busy events.

As Ray Culbertson changed to 1.0 the radius parameter in his analysis program and re-run it, his jaw dropped. The dijet mass distribution now showed a towering three-bin peak. Even more impressive was the fact that the background prediction followed very precisely the data in all the other bins. It was hard to ignore the possibility that there was a resonance sitting on top of a smooth continuum! Culbertson of course well knew that it was not easy to defend his choice of a wide jet radius: the signal was not there with the standard radius. That was discomforting. On the other hand, the signal he found with the larger radius really did look like a new particle.

Ray and collaborators produced a CDF note describing the anomalous effect and all the checks and studies they had been able to perform.

They tried different jet energy correction algorithms, alternative b-tagging tools, and all possible jet radii. They dissected the sample by studying individually every event in the peak, trying to figure out whether it was a good idea to use wide jets in each of them. Unfortunately, a clear conclusion on the origin of the peak could not be reached. Some features appeared to confirm its exotic nature, whereas others seemed to indicate that it was due to a statistical fluctuation. The analysts had no choice but to rest their case: their dijet bump remained an unsolved anomaly. It could not be called a discovery, nor be disproved as a new physics signal. It would be another business for Run 2 to straighten out.

Chapter 9

Preon Dreams

In this chapter we deal with one of the most intriguing and exciting among the several anomalies that arose in the analysis of Run 1 data. One which lent itself to an immediate, startling interpretation: maybe quarks are themselves composite objects after all! The possibility of such a revolutionary discovery sucked the CDF collaboration into the center of the first media storm of its career. The coverage of the story by the press was comparable in intensity to the news on the top discovery, which had been announced only a few months earlier. But, in this case, the management of the experiment was caught entirely unprepared. The unanticipated and uncoordinated interviews that science reporters obtained directly from members of the collaboration, as well as the uncontrolled claims artfully inserted in some of the resulting articles, generated a good deal of friction within the experiment.

Tortoises All the Way Down

The idea that quarks are composed systems made of smaller entities generically dubbed *preons* (from the concept of "pre-quark") was not new in 1995. The suggestion had been made already two decades earlier, in a 1974 article by Jogesh Pati and Abdus Salam. It might be seen as a natural reaction to the very concept of "elementarity": the existence of entities that have no spatial extension, as point-like and structureless as the standard model quarks and leptons, is something totally alien to our understanding.

Hence the natural appeal of the suggestion that those bodies, which we call elementary, be themselves composed of smaller objects.

Theorists saw preons as a tool to try and address many of the riddles to which the standard model offered no solution. What makes quarks different from leptons? Maybe the answer was that they were composed of smaller building blocks arranged in different ways. And why three generations of fermions, and not just one, or four, or 12? This again could hopefully find an elegant, simple explanation. Preons could also be hoped to, one day, allow a direct calculation of the many free parameters of the standard model, such as fermion masses and coupling strengths. Finally, preons might even conceivably provide alternatives to the Higgs mechanism in the explanation of electroweak symmetry.

Before the advent of the Spp̄S and Tevatron colliders, the preon hypothesis was not directly testable with particle collisions. The dynamics of quarks and gluons are governed by QCD, a theory which back then was not understood well enough to be the basis of precise tests of quark substructure. QCD was still in need of experimental verification: it was not solid ground on which to stand up and look beyond. Also, the available center-of-mass energy of the collisions studied in the 1970s was barely sufficient to "resolve" quarks within the proton. Electron–positron collisions had just been energetic enough to enable the observation of the first low-energy hadronic jets (which were a manifestation of the creation of quark–antiquark pairs) and the discovery of the gluon as an occasional third jet in those events. As for proton collisions with fixed targets, the released energy was sufficient to probe the structure of the proton, but it did not allow one to go deeper. In contrast, indirect tests of a lepton substructure could already be achieved through a detailed study of the magnetic properties of electrons and muons. Taken at face value, the good agreement of those properties with electroweak model predictions implied that if leptons were composite, their constituents had to have very large masses, above a few hundred GeV.

To most theorists, the study of how preons could make up quarks and gluons initially looked like an unexciting prospect, as there were no experimental hints to use as a guideline. In addition, there was the risk of seeing their work on the matter ignored for many years to come, due to the lack of a direct testability of the results. It was only during the 1980s that the idea of preons acquired the largest following. Theorists started to combine in various ways the idea of a grand unification of all forces with that of the existence of simpler elementary constituents of quarks and leptons. Preons were also studied in the context of supersymmetric models and other complex concoctions, such as "superstring" models and "technicolor" theories. The publication of theoretical papers containing the word "preon" in the title peaked in 1985. One of those works was titled "Preons and their implications for the next-generation accelerators" and was authored by Jogesh Pati. Again, the inventor of quark constituents provided a detailed summary of the situation and also expressed his views on the concept of elementarity of matter:

> "If quarks and leptons are composite of "preons," one may, in general, permit the possibility that preons themselves may be composites of more elementary objects —"pre-preons" — and so on. We believe that this chain of increasing elementarity will end, but only when one reaches a stage which is definitely economical, elegant and somehow "unique." After all, that is the primary goal of all searches for elementarity."

Here we see how deeply rooted is in the mind of physicists the *lex parsimoniae* outlined by William of Ockham seven centuries ago: economy is of paramount importance, so much so that it defines the characteristics of the laws of physics. We expect the world to be governed by simple and elegant rules and we would not understand the beauty of the construction if the latter lacked economy. Hence, the idea of facing a regression ad infinitum in the quest for the truly elementary constituents of matter is not to the physicist's liking. One is reminded of the funny story recounted in Stephen Hawking's 1988 book "A brief history of time," where according to a little old lady arguing with the speaker at an astronomy conference, the world is supported on the back of a tortoise. Asked by the scientist what is the tortoise supposed to be standing on, the lady replies that it is "turtles all the way down."

The theory of preonic constituents within quarks and leptons offered a clue of the energy scale at which those constituents would show up, as values of the order of 2–3 TeV were favored in Pati's view: quoting again from his paper,

"Based on our considerations of dynamical symmetry breaking through preons and the assumption of universal size for all quarks and leptons, we are led to predict that quarks and leptons have an inverse size of the order of 2–3 TeV."

Above, Pati is implicitly using the existence of an inverse relationship between energy and distance scales. Because of that, higher-energy collisions probe smaller distances. If Pati's prediction were true, measurable effects might be expected to become visible as one studied processes releasing energies corresponding to half of that inverse size, i.e., above 1 TeV or so. A further quantitative prediction was offered for the cross-section of high-energy collisions between hadrons. Pati suggested that the consequence of the existence of preons in quarks would "show in deep inelastic ep or pp scatterings [...]," since it "would lead to a 50% departure in the cross sections [...]" as one reached the energy scale corresponding to the inner structure of quarks. That prediction was at the basis of the excitement that jets of high energy would cause in CDF ten years later.

Jet Energy Spectra for Dummies

The high jet-E_T excess, as the anomalous tentative signal of quark compositeness was going to be called by CDF members ever since 1995, surfaced in a study that had no special ambitions to probe new physics models. This was a measurement of the properties of quantum chromodynamics, carried out through the counting of energetic jets. The idea of the study is very simple: you take all events where you see an energetic jet and use them to construct a transverse energy spectrum: a histogram obtained by distributing the data into bins of different jet E_T.

Here, it is useful to recall that high-energy physicists are mostly interested in the component of the motion of final-state particles *transverse* to the beams direction, as this carries information about the strength of the interaction withstood by the colliding partons. The longitudinal component of the motion is much less informative, as it receives arbitrarily large and undetermined contributions from the uneven combined momentum of the two partons colliding in the initial state. This is nothing but the law of conservation of linear momentum. If a parton with a momentum of, say, 300 GeV along the proton beam direction hits another parton traveling in the opposite direction with a momentum of 10 GeV, the center of mass of the hard collision moves along the proton beam direction with a total momentum of 290 GeV. All the products of that collision are thus going to be boosted in the same direction. Yet, the total energy release is not very large, because most of the kinetic energy of the colliding partons is converted into longitudinal momentum of the final-state objects rather than being made available for the creation of new massive states.

In light of the above, hadron collider physicists often even forget the adjective "transverse" when they talk of transverse energy of observed particles or jets. It is considered implicit in all but very few specific cases and its omission usually causes no confusion. Another thing worth pointing out is that in the case of an energetic jet, the concepts of energy and momentum are interchangeable, as they only differ when the mass of the system is not negligible, while it usually is. That is why one loosely talks of transverse energy ignoring the fact that energy is a *scalar* quantity, i.e., a number with no associated direction, and not a vector like momentum, which has three spatial components: "transverse energy" is then a synonym of "transverse momentum."

Even before you fill your jet E_T histogram, you know how it is going to look like. The number of jets will rapidly decrease as E_T increases. That correct expectation is due to the peculiarity of parton distribution functions mentioned in Chapter 2. If you take a snapshot of a proton accelerated to very high energy, you are going to see in it many quarks and gluons, most of them carrying only a very small fraction of the total proton energy. Hence, when one of those partons collides with a parton in the antiproton, the center-of-mass energy of the pair is quite likely to be a small fraction of the total energy carried by the proton–antiproton system. Very hard collisions, in other words, are rare. Correspondingly, high-energy jets are more infrequent, and more interesting, than low-energy ones. Transverse! I meant transverse energy, shoot.

Then the question arises of what data to use as input. The online data acquisition system works by selecting only the most significant collisions among the hundreds of thousands that take place every second in the core of the detector, and for events with jets, "most significant" is a synonym of "with jets of highest E_T." So, you might imagine that the data you need are those coming from a trigger selecting events with jets of high E_T. "Below 100 GeV? Off to the trash bin"; "Above 100 GeV? Please follow this way to the data storage system." And, in fact, if you remember the description of the CDF trigger in Chapter 2, you might recall that its logic was to "dig deeper," collecting all the events which could shed light on phenomena that lower-energy colliders had no chance to study in the past, and discarding the others. However, that was only a useful simplification. The trigger of CDF was a quite complicated thing because it made every possible effort to not burn any bridges, as the collection of a single dataset with jets above a fixed threshold would have undoubtedly done. Instead, triggers with looser energy thresholds were also at work, to avoid entirely losing the information about those less dramatic but still useful collisions: there were jet triggers with thresholds of 70, 50, and even only 20 GeV, a very loose requirement. The resulting datasets allowed CDF to perform calibrations, study backgrounds, and improve the detector modeling. But they were *prescaled*: only one every 10, or 100, or even 1000 events fulfilling the trigger requirements was actually sent to the output data stream. To explain how this worked, we may consider a numerical example not too far off the real situation.

Let us imagine that you study the average rate of jet events at a given instantaneous luminosity, and you observe rates of 2000, 100, 20, and 5 events per second as the E_T threshold is set at 20, 50, 70, and 100 GeV, respectively. You are interested in jet events and you would be glad to get all of them, yet a frowning Hans Jensen, the thick-bearded leader of the Trigger group, tells you that you are allowed to store no more than 10 jet events per second. This is because other triggers, which select precious events containing electrons, muons, or other striking signatures, must also have a chance to write those events to tape, and the total output capability of the data acquisition system is of no more than 50 events a second!

So, you decide to take all of the five events per second with jets above 100 GeV, which are by far the most interesting ones. At this point, your bandwidth budget allows you to record five more events per second, while events with jets still requiring a decision pour in at the terrifying rate of 2000 per second (a jet passing the 50, 70, or 100 GeV threshold of course also passes the 20-GeV one, so 2000 per second is both the rate of the

20-GeV trigger *and* the total jet rate). What to do? The solution is that you use a *prescale*, a device which saves one every N events passing a given trigger threshold and dumps the other $N - 1$. You can, for instance, set the prescale factor of the Jet-70 trigger at 1-in-10 ($N = 10$): this will ensure that a rate of only two events per second will be written to tape, out of the 20 that pass the 70-GeV threshold. You can also set the prescale factor of the Jet-50 trigger at $N = 100$, or 1-in-100, so out of the 100 events per second with a jet above 50 GeV, only one gets written to tape; and finally, you can set the prescale factor of the Jet-20 trigger at $N = 1000$, for 1-in-1000, which ensures that of the 2000 such events per second only two get actually stored.

Let us take stock: every second, the above arrangement of trigger thresholds and prescale factors yields on average five 100-GeV events, two 70-GeV events, one 50-GeV event, and two 20-GeV events, for a total of 10 events per second. There you have a recipe for using effectively the bandwidth granted by Jensen, while ensuring that you collect all the highest energy jet events, as well as a *representative sample* of jets with smaller energies. In a 100,000-second store (a typical one), you get 200,000 events passing the Jet-20 trigger, 100,000 passing the Jet-50 trigger, and 200,000 passing the Jet-70 trigger. Those prescaled datasets are large enough to allow for the study of the characteristics of jets in each energy interval with a similar statistical power.

Note that you could have also gotten away with setting a single prescale factor of 400 on jets above the 20-GeV threshold and ignored altogether the Jet-50 and Jet-70 triggers, arguing that the Jet-20 selection automatically also collects jets that pass those higher thresholds. This would work in terms of bandwidth, as the initial rate of 2000 events per second prescaled by a factor of 400 would still yield the five events per second which you are allowed to store in addition to the five Jet-100 ones. However, the result of such alternative approach would be quite unbalanced: in the course of the 100,000-second store, the collected sample of jets with E_T above 20 GeV would contain half a million events, but of these only 25,000 and 5000 would have jets above 50 and 70 GeV, respectively. Such a choice would enable you to measure very well the properties of low-energy jet events, but any study involving jets in the 50–100-GeV region would then suffer from very low statistics.

Once you have your data ready, you can proceed to construct the energy histogram. Besides factoring in the different prescale factors that the different jet datasets have been subjected to, you must also account for subtle effects that modify the observed rate, such as detector inefficiencies or jet reconstruction failures. And then you must also calibrate the energy measurement of the observed jets, using the precise knowledge of how the detector responds to particles of different energy in different admixtures. This calibration allows you to estimate the true parton energy from the observed jet energy, so that you can put each event in the E_T bin it really belongs to. The above tasks demand a long and painstaking study. At the end of it, though, you receive a well-earned prize: you can finally compare your result to theoretical models. This is the fun part, as you can now judge whether one model is better than another at reproducing the behavior of proton–antiproton collisions.

Because of the connection between parton distribution functions and the energy with which partons emerge from the collision, the shape of the histogram is a direct way to test our knowledge of the PDF and of the proton structure. There exist many different parameterizations of the proton PDF on the market: the different models are produced by independent groups of phenomenologists. They dream up some functional forms for the PDF (the model). Then they determine the model parameters by fitting together the experimental input from all available measurements sensitive to the way quarks and gluons share the energy of their parent. Those fits are periodically updated to include new measurements or to reassess the relative importance of specific datasets.

The final result of a study of jet rates as a function of jet E_T is usually a graph where experimental data are overlaid to several model predictions. For instance, this could show that the PDF model produced by group A is a good one, as it matches the data well, whereas the model proposed by group B fails to reproduce the observed distribution in at least a part of the spectrum. When this information becomes available, both groups A and B rush to insert the new data into their analysis programs. The parameters of group A's model are then likely to change very little after the inclusion, whereas the parameters of B's model suffer a larger shift, reflecting the pull that the data exert on B's model parameters in the region where a discrepancy is observed. As a result, group B may decide to tweak the functions they use in the model, to give them more freedom to interpolate all available experimental data. The improved models are then made available to the experiment, providing a basis for more detailed tests as additional data are collected or new analyses are performed. This back and forth between model fitters and data analysts could be viewed as an idle game, but it actually is good science: only through a precise knowledge of the PDF may the experiment hope to observe the effect of some new physics processes in their data, as the PDF are a crucial input in any Monte Carlo calculation of standard model backgrounds.

The Run 1A Measurement

The physicists who took on the task of producing a measurement of the jet E_T spectrum using Run 1A data were Brenna Flaugher and Anwar Bhatti. Anwar was, at the time, a post-doctoral scientist with the Rockefeller University group. He had joined CDF toward the end of 1991, when Steve Kuhlmann suggested he work on the inclusive jet E_T cross-section. In addition to being a co-convener of the QCD group in CDF, Kuhlmann belonged to the CTEQ team, a group of theorists and experimentalists who authored one of the most trusted PDF models. He persuaded Anwar by arguing that it would be the most important measurement Anwar was going to do in his whole career — with hindsight, a visionary statement. Brenna instead had already measured the jet E_T spectrum with Run 0 data, publishing her result in 1992; now she was naturally curious to see what she would get with the five times larger dataset collected in Run 1A. In an academic world dominated by the publish-or-perish rule this was also a cost-effective choice, as she had acquired experience with the analysis tools needed for the measurement: a new publication of the same physics process would take less time and effort to be produced. But, in addition, it promised to be interesting, for two reasons.

Because of the record energy of the Tevatron, CDF data probed a region of parton kinematics where past experiments had provided little or no information. The higher statistics of Run 1A allowed Brenna to study a wider energy range than the previous measurement, pushing the analysis to jets with transverse energy up to and above 400 GeV. This is because of the vanishing of PDF at the highest momentum fractions: a larger integrated luminosity extends the range of studied energies even if the proton–antiproton collision energy stays the same, as the rarity of the highest energy subprocesses demands higher "illumination" to make them out, as suggested by the speleologist analogy discussed in Chapter 2.

The second reason for repeating the measurement was an unspoken one. Run 0 data had provided a great confirmation of QCD, offering a beautiful agreement of observed and predicted jet rates over a wide range of subprocess energies. But it had also provided some feeble hints that the least well-known region of the spectrum, its high-end tail, could be hiding some surprise. A small excess of very high-E_T jets over theoretical predictions had, in fact, been seen. The predictions came from QCD calculations of the parton scattering processes, coupled to the available models of parton distribution functions. The effect was small from a statistical standpoint; but to Brenna it was a curiosity waiting to be satiated. Would the increased statistics of Run 1A confirm or disprove the faint Run 0 excess?

A measurement of the scattering of partons off one another can suddenly turn into an extremely intriguing investigation, if the result defies one's expectation. The very high-energy end of the jet spectrum corresponds to reactions that involve the largest energy exchange between the partons. As higher-energy collisions probe smaller distances, those collisions are the ones which contain more detailed information about the structure of matter at its smallest scale. At center-of-mass energies of a few tens of GeV, quarks and gluons become evident within the proton; at center-of-mass energies of the order of a TeV maybe preons start to make themselves felt, too, if they exist!

In addition to being sensitive to the existence of a quark or gluon substructure, the high-E_T jet events have, in principle, another way to manifest the presence of new physics. Several theoretical ideas that extend the standard model, in fact, hypothesize the existence of new neutral vector bosons, generically called *Z-prime* and labeled Z'. Except for their large mass, Z' particles would have properties similar to those of the well-studied Z boson, which is known to decay 70% of the time into quark–antiquark pairs, and only 7% of the time into electron or muon pairs (the remaining decays yield neutrinos or tauons). The existence of a 1-TeV Z' boson could become evident by observing an excess of, say, one or two dozen events with pairs of jets of transverse energy in the 400–500-GeV range. Due to the 10 times higher chance of the hadronic final state and the higher solid-angle acceptance of the CDF detector to jets than electrons or muons, this would happen well before one would be able to detect a Z' through its alternative and cleaner decay to electron or muon pairs. The leptonic final states would be way more striking and background-free, but they might not have a chance to appear in time.

As discussed above, the construction of a jet E_T histogram from inclusive jet data may not sound too complicated. However, producing the "raw" histogram is only the first step in a careful analysis. Anwar and Brenna had to operate a series of corrections in order to obtain an accurate estimate of the true jet cross section as a function of energy from the observed rates of jet events. The most important of those corrections was a careful calibration of the transverse energy of jets: due to the strong dependence of event rate on jet E_T, even a small underestimate or overestimate of the latter results in a very large bias in the former.

The analyzers presented with continuity the progress of their analysis at the fortnightly meetings of the QCD group, receiving support and suggestions from the regular attendees. Unlike the Top group, this was a relaxed and largely non-competitive environment where collaborators did what they were supposed to do: collaborate! Indeed, several colleagues provided a significant contribution to the measurement: the most active along with Kuhlmann were Rob Plunkett, Rob Harris, Eve Kovacs, Liz Buckley, and Steve Geer. All of them were performing similar analyses which used in part the same tools — QCD predictions, PDF sets, jet reconstruction and calibration methods. In addition to these experts, the QCD group included colleagues who would later be designed as the godparents of the analysis: Steve Behrends, Marjorie Shapiro, and Tom Devlin. The jet E_T cross-section measurement was one of the most important measurements that the QCD group planned to produce from the analysis of Run 1A data, so it was natural that a constructive collaborative effort would take place.

As soon as Brenna and Anwar put together all the calibrations and corrections to the raw histogram and produced what looked like a close-to-final result, they discovered that the spectrum really did diverge at its high-E_T end from QCD calculations, in a way which looked significant. For jet E_T below 200 GeV the data agreed very well with theory curves, but

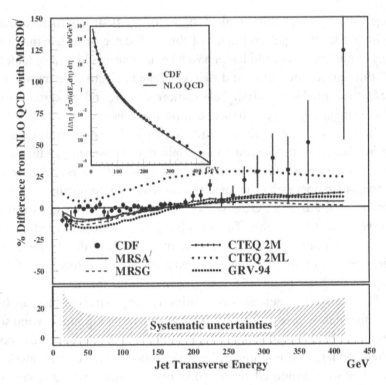

Figure 1. Comparison of the jet transverse energy spectrum obtained by CDF from Run 1A data with QCD predictions employing different PDF sets. The upper panel shows the fractional difference between experimental data and predictions of the MRSD0' group as points with uncertainty bars; the predictions of other PDF models are shown by different hatched curves. The lower panel shows the extent to which systematic uncertainties may change the fractional difference. In the upper left inset is shown the original distribution, evidencing the good agreement of the jet production rate over several orders of magnitude, as well as the size of the departure at the right end of the spectrum. Reprinted with permission from *Phys. Rev. Lett.* 77 (1996) 438.

it departed from predictions above that value. If one used the fashionable *MRSD0'* PDF model as a baseline, for instance, the data exceeded the predictions by 30–50% at 300 GeV and by a startling 120% above 390 GeV; the use of other PDF models yielded similar results (see Figure 1). The analysis withstood a deep level of scrutiny within the QCD working group, and a long set of questions and requests were addressed before Brenna and Anwar could manage to present a blessing talk. Once that was past them, the result was public: it could be shown at conferences, and publications could include the graphs approved for public consumption.

One good occasion to discuss the new observations by CDF came in the late spring of 1995, when Anwar Bhatti gave a talk titled "Inclusive Jet Production at the Tevatron" at the 10th topical workshop on Hadron Collider Physics. In his presentation, Anwar represented both CDF and DZERO, so he discussed in succession the measurements of the jet E_T spectrum that each collaboration had produced with Run 1A data. DZERO also observed an increase in their jet rate at high energy, but their measurement had larger uncertainties. Statistical uncertainties were larger as DZERO had analyzed only two-thirds of the available collisions; and systematic uncertainties were dominated by the insufficient knowledge of the jet energy scale, which translated into a 35% uncertainty on the cross section. The DZERO result was thus compatible both with the CDF one and with theoretical models: it did not provide further insights into the origin of the CDF anomaly.

Along with the jet E_T cross section measurement, Anwar reported in his talk the results of two additional analyses which were, in principle, sensitive to the same sources of deviations, be it new physics or incorrect theoretical models. One was a new dijet mass spectrum measurement performed by Rob Harris, who had managed to include not only the Run 1A data, but also a good portion of the fresh new Run 1B data. The second analysis, produced by Steve Geer and Liz Buckley, was a measurement of the cross section of jet events as a function of the sum of the transverse energy of all jets, a quantity of course highly correlated with the E_T of the most energetic jet. Indeed, both the dijet mass spectrum and the total transverse energy spectrum showed high-tail excesses similar to those of the inclusive jet E_T analysis: the effect was clearly systematic, and it did not seem to depend on analysis details.

As usual in similar situations, the CDF result was met with excitement mixed with skepticism. Anwar had made it clear in his discussion that a possible explanation of the observed effects was the incorrect modeling of the probability of high-momentum partons in the PDF models. But he also, of course, pointed out that the size of the increase in the rate of high-E_T jets matched what one would expect from a *compositeness scale* (the inverse size of parton constituents) of 1–2 TeV. This was encouraging to those who wanted to believe in the reality of the preon hypothesis.

Anwar also explained that a new analysis was starting which would study the angular distribution of high-E_T jets. That would cast more light on the nature of the excess, since a quark compositeness signal yields much more central jets than energetic gluon–quark scatterings.

The jet E_T spectrum was the topic of extensive discussions at the particle physics conferences that took place in the summer of 1995. The ubiquitous issue was an objective evaluation of the statistical significance of the discrepancy. This was clearly the critical point on which to concentrate the efforts of the group before a scientific article could be produced. The significance was initially studied by Tom Devlin, a professor at Rutgers University, and independently by Kuhlmann. The two physicists were very familiar with statistical calculations, but they had different opinions on the way to account for systematic uncertainties; due to their different approaches, it proved difficult to compare their results. Anwar did not feel comfortable with the situation; in particular, he did not agree with some of the choices of Devlin. Also, Kuhlmann's evaluation of the effect returned a probability of 0.03% of observing data at least as discrepant with the model. This looked really small. Anwar decided to get on board for a third-party evaluation his Rockefeller group colleague Luc Demortier, who was one of the CDF members most knowledgeable in statistical techniques.

It might appear strange that a typical problem in data analysis at a particle physics experiment would cause so much trouble. In fact, the assessment of the significance of a departure of data from a model at the tail of a spectrum is a very common task in particle searches, although not at all a trivial one. The tails of mass or energy distributions are usually the focus of experimentalists because they represent the "frontier" of a measurement, the region where data are produced by the rarest reactions. However, statistics always provides you with the chance of cooking up quite different answers to the same question, depending on subtle arbitrary choices you make (sometimes unknowingly) in working them out. To a physicist, this is quite unsettling!

Kolmo-Who?

There were various issues in the determination of the statistical significance of the high-E_T excess. First of all, the proper accounting and modeling of all known systematic effects capable of affecting the predicted distribution was extremely complicated. One could imagine several causes of shifts and tweaks: ill-known hadronization properties of jets at low or high energy, nonlinearity of the response of the calorimeter, failures in the QCD modeling of the reactions, etcetera. Those sources could both modify the shape of the distribution — i.e., the relative rate of events at low or high E_T — and the overall rate of jet events. Each effect needed to be properly studied to assess its impact on the measured jet rates at different jet E_T. Only then could one account for all the systematics in the calculation of the test statistic and finally obtain a p-value. But even at that point there remained a quandary, one to which statistical textbooks could not indicate a univocal solution.

The spectrum was constructed with tens of thousands of jet events, and the agreement between model and data was excellent below 200 GeV, where most of the events fell. That was not at all surprising, as the model to which the data were compared was the result of fitting data from many previous experiments which studied in detail that same energy range. The disagreement only appeared in the *terra incognita* of very high energies, but there the number of data events was of course much smaller. In such a situation, if one performs a hypothesis test on the whole spectrum, the test will likely confirm the *null hypothesis* (i.e., the agreement of data and tested QCD model), as the statistical power of the data is highest at low E_T, where the agreement with the model is good. Only a test restricted to the high-end tail of the spectrum has a chance to return a low p-value of the null hypothesis. Hence, an objective quantification of the probability of the effect is impossible: the result strongly depends on the range of the spectrum used in the test. And physicists asked for an opinion will no doubt have one in such situations. Devlin argued that one should pick as tested region of the spectrum the one producing the highest discrepancy; Anwar disagreed.

Demortier attacked the problem by producing several different statistical tests, using different prescriptions for the test statistic of choice and studying how the results could vary as a function of the arbitrary cutoffs defining the investigated jet E_T region. The rationale of investigating a wide spectrum of choices was clear: one wanted to ascertain how robust could be the conclusions drawn from the data. It would have been questionable to rely on one particular statistical test without a proper justification for its choice, hence Demortier tried the Kolmogorov–Smirnov test, the Smirnov–Cramer–Von Mises test, and the Anderson–Darling test. Each of those prescriptions had its points of strength and its weaknesses. Demortier, however, found that the Anderson–Darling test statistic provided the best *a priori* sensitivity to deviations of the distribution in its tail. On November 20, 1995 he documented his studies in CDF note 3419, titled "Assessing the significance of a deviation in the tail of a distribution."

A few weeks later, the matter of what prescription to use for the quantification of the *p*-value was decided during an open discussion at a collaboration meeting. The method proposed in CDF note 3419 was accepted, and the lower end of the energy range considered in the test was somewhat conservatively set at $E_T = 200$ GeV. As for the choice of the test statistic, it eventually fell on the Anderson–Darling one. However, the winning argument was not based on Demortier's studies of discriminating power, but rather on a point made by Tom Devlin:

> "All of these tests are rather unfamiliar to PRL readers. We should pick a test with an Anglo-Saxon name!"

At the end of the day, using defensible choices for the PDF set and related parameters as the null hypothesis, the *p*-value of the data was computed to be of the order of 1%. One percent is, in absolute terms, a small number: for instance, if you are studying the correlation of drinking habits with the incidence of a particular kind of cancer and you find that there is a 1% chance of observing your data if the two things are uncorrelated, you happily write in your paper that you have proved a link between them. In fact, in medical research, a 1-in-100 observation is considered significant. However, 1% is definitely not small if you want to base on it a claim of new physics; once converted in a number of standard deviations, it amounts to a less than "three-sigma" effect. Was the high-E_T anomaly not so surprising after all? It was clearly an oddity of the data, but the majority of the CDF collaboration felt that it was not enough for any well-grounded claim; and yet it was one on which intriguing speculations could be based. It seemed reasonable, given the high relevance of the alternative interpretation of the effect, to include its discussion in a publication describing the measurement.

The different feelings on the overall message of the study started to weigh in as the publication describing the jet E_T cross-section measurement got finalized. Kuhlmann did not agree with some of the details in the statistical analysis performed by Demortier; he argued that the anomalous effect was a more significant one than what a 1% p-value implied. Yet, Marjorie Shapiro, who chaired the godparent committee in charge of the analysis review, believed that the chosen conservative approach was more in line with the feelings of the collaboration. The emphasis had to be put on the explanation of the level of understanding of the systematic effects which had the power to affect the agreement between data and models, rather than on the overall smallness of the p-value. As Demortier's statistical analysis was the better motivated and documented, and it also returned a less striking disagreement with the null hypothesis, there was no way to strike a compromise.

The argument eventually led Kuhlmann to ask that his name be removed from the authors list of the publication, as the paper got ready to be submitted. At that time, he did not believe it was so big a deal: he had only meant his gesture to signal his minor disagreement with the way the effect was reported in the article. Nonetheless, Kuhlmann's retraction eventually generated some turbulence. There were two sides of this. One was the unspoken law of consensus, according to which the CDF collaboration had to speak with a single voice, as dissent would be interpreted from the outside as lack of confidence in the published results. The absence of one signature in 500 or so was a step away from that policy, albeit a small one. The other side of it had to do with the media storm which was preparing at the horizon. Several CDF members, who disagreed with the publication of the allegedly ill-understood E_T spectrum but did not take the stand of refusing to sign the article, ended up regretting it when they realized how controversial the matter had become — why, Kuhlmann had taken his name off, and they had not been smart enough to do the same....

The writing of the article had already started in the early summer of 1995. The descriptions of the motivation of the study, the experimental apparatus, and the technique used to extract the jet energy spectrum were not an issue. What was critical was to explain convincingly the detailed studies of systematic uncertainties, and the statistical procedure which estimated the consistency of the data with theory predictions. Then there were other connected issues: how much weight to give in the paper to a discussion of the new physics alternatives and their success in providing an explanation of the observed excess; where to place that discussion in the text; and how much of that information to propagate to the abstract and the conclusions section of the paper.

Anwar wrote a first draft of the article, and later Brenna took over the task of implementing changes. When the draft was finally circulated within the collaboration, in order to get a first round of suggestions and feedback as was normal practice for any planned publication, authors and godparents got their mailboxes flooded with reactions of all sorts. A very small minority of the respondents objected altogether to publishing the results, on the basis that the high-E_T disagreement with QCD was evidence that the data were not sufficiently well understood. Many others wanted the paper to describe the measurement without any speculation of the potential new physics sources of the high-E_T effect. Some even objected to a detailed discussion of the evaluation of statistical significance of the anomaly, arguing that any quantitative treatment would give too much emphasis to the effect in the article.

It was difficult to converge to an agreed-upon text also because the CDF collaborators looked like a bunch of doctors, each carrying a different opinion on how to treat a patient with complex symptoms. The observed effect was of course just a statistical fluctuation; it was evidently a clear deficiency in the models to which the data were compared; it was probably an underestimate of jet energy scale systematics; it indicated that the QCD coupling constant (a parameter which determines the rate of QCD processes, and which is in fact not constant, but varies with energy) needed to be evaluated at a different energy scale; it was a detector effect combined to poorly known details of hadronization; it was a combination of a statistical fluctuation with systematic effects; and so on. The large majority of the collaboration members seemed to have their favorite no-new-physics explanation for the anomaly. Only a narrow minority wanted to emphasize the new physics explanations in the text.

Amidst this confusion, a few QCD experts had well-founded reasons to believe they knew the correct answer. One of them was Kuhlmann. He had already been studying for a while how some of the old results from fixed-target experiments, which were routinely included in PDF fits, were at odds with the rest of the available data. Their removal from the fits, or a scaling up of their uncertainty, allowed for a considerable extra freedom of the PDF model. In particular, it was possible to hypothesize a higher probability of finding energetic gluons in the proton. That would increase significantly the rate of energetic quark–gluon collisions, boosting up the cross section at high jet E_T and softening the disagreement with the CDF data. Unfortunately, Kuhlmann's studies with the CTEQ group were not quite ready yet to be published at the time of writing of the CDF article, so they could not be cited. There was no choice but to publish the anomaly without that natural explanation.

Because of the strong and incoherent reaction of her colleagues, it proved quite hard for Brenna to converge to a text that could meet the consensus of the majority of the collaboration. The detailed wording of every sentence withstood dozens of iterations, often getting into the annoying oscillatory mode between two or three different versions of the same paragraph. But eventually a middle point was found. And, since theorists would have certainly tried to fit the CDF excess to compositeness models anyway, but they lacked all the internal information required to obtain a precise result, CDF in the end decided to include in the paper the result of that exercise, which determined the compositeness scale best matching to the high-E_T excess.

It was January 25, 1996 when the paper was submitted to PRL and to the Cornell preprint ArXiv. The conclusions of the article did contain the result of the compositeness fit, but followed it with a very down-to-earth statement:

> "[u]ntil a realistic method for representing the theoretical uncertainties from higher-order QCD corrections and from the PDFs is found, any claim about the presence or absence of new physics is indefensible."

The Media Storm

At the time of submission of the paper, the particle physics community had already digested the CDF result. Public debates on whether to call agreement or discrepancy the comparison of the CDF spectrum to its DZERO counterpart had been losing strength. Both collaborations were more interested in improving the statistical power of their measurements by using the five times larger Run 1B datasets which had recently been made available to analyzers. The QCD group in CDF was also in the process of producing a precise measurement of the angular distribution of jet pairs. That measurement used the same events which populated the highest E_T bins of the inclusive jet E_T distribution. As noted earlier, the angular distribution offered a powerful test of the preon hypothesis.

In the course of January 1996 *Science* magazine conducted several interviews and collected material for a publication of the story. The *Science* reporter was James Glanz, who had a Ph.D. in astrophysics: this was close to the best possible scenario that CDF could hope for, as the article was guaranteed to contain neither sensationalistic claims nor incorrect or deceiving statements. And, indeed, it did not. However, *Science* had a policy of "leaking out" to local newspapers the most interesting articles they were about to publish, just before the magazine got distributed. And there was the rub. Would local newspapers match the high standards of the flawless text produced by Glanz? Not a chance.

In the February 9, 1996 issue of *Science*, James Glanz titles openmindedly: "Collisions Hint That Quarks Might Not Be Indivisible." His article is quite good and balanced, and it explains the physics well, without making unsupported claims. The article also contains several quotes from CDF members, starting from the spokespersons:

> "This is just the sort of effect you would see" says CDF-co-spokesperson William Carithers, "if quarks were not fundamental particles but had some sort of internal structure."

And then Giorgio Bellettini:

> "As these events began to accumulate [...] "a fierce fight" broke out within the collaboration over how to gauge the small chance that systematic experimental errors could explain the results. The researchers made exhaustive tests of the possibility that a "conspiracy" of random or systematic errors might be fooling them [...]. Finally, he says, the collaboration reached a consensus that the excess had to be real."

The article closes with a quote from Steve Geer, the other co-convener of the QCD group, who suggests what will only later be accepted as the correct solution:

> "Steve Geer, a CDF team member at Fermilab, describes the most dramatic possibility: "It might mean that, just as in Rutherford's atom, there's a hard center" lurking inside the quarks, as some speculative theories suggest. But Geer points out that several other explanations might account for the measurements. The more mundane possibility, he says, has to do with how momentum is parceled out among the components of a speeding proton. The hardest collisions occur when two quarks that happen to carry a high fraction of each proton's momentum meet head-on. But the mass-less gluons can carry momentum as well. So if, say, QCD underestimates how often gluons carry a high fraction of the momentum, then the quarks they encounter could suffer an unexpected number of violent collisions, and "we could end up with more energetic jets than expected.""

Overall, the *Science* article does not appear to be damaging to the CDF collaboration: no strong claims, only a report of work in progress. But in 1996 an influential minority of the CDF members was uneasy with any sort of media coverage of their work. A few of them were quite frightened that the lack of control over what could be perceived from the outside might damage their reputation and worsen their chances of getting money and grants from the funding agencies. This paranoid feeling had its origin in a piece that Malcolm Browne had published on the *New York Times* on January 5, 1993. The article titled "315 Physicists Report Failure in Search for Supersymmetry" and was built around the rather unremarkable news that a publication had been sent by CDF to *Physical Review Letters* to report lower mass limits in a search for squarks and gluinos. Browne had insisted on reporting the point of view of critics of "big science." To let you taste the flavor of the piece, here are a few clips:

> "[...] despite this arsenal of brains and technological brawn assembled at the Fermilab accelerator laboratory, the participants have failed to find their quarry, a disagreeable reminder that as science gets harder, even Herculean efforts do not guarantee success." [...] "Some regard such

failures as proof that high-energy physics, one of the biggest avenues of big science, is fast approaching a dead end." [...] "Failed experiments of such a grandiose scale offer easy marks to critics who contend that 'big science' produces too meager a crop of knowledge for what it costs." [...] "all these projects soak up more human talent and more public money than society should permit, skeptics argue."

To his credit, Browne concluded the article with a nice quote by Wolfgang Panofsky, the director-emeritus of the SLAC laboratories:

"If certain answers crucial to man's understanding of nature can be obtained only by large effort, is that sufficient reason for not seeking such answers?"

Alas, Panofsky's quote was not nearly enough repairation of the damage of the previous ones. The potential danger of similar press coverage was well understood by some CDF members, who remembered vividly how a few months after the publication of Browne's article the US Congress had cut the funding to the project of building a Superconducting Super Collider (SSC) in Waco, TX. This was a hadron collider meant to reach the energy of 40 TeV with proton–proton collisions. The demise of the SSC was a bad blow to the US plan of research in frontier particle physics, and shattered the hopes of many who had worked hard to design the detector and study the physics potential of that facility.

Of all the quotes in Glanz's article reported above, the one by Bellettini was badly received by some collaborators. It spoke of fights within the collaboration and of a "real effect"; and it came from a spokesperson, which was an aggravating clause. Worse still, as the *Science* paper hits the press Bellettini was also featured in a *National Broadcast Radio* interview, where he explained the matter in his usually colorful, enthusiastic way. Bellettini has always been a true enthusiast of particle physics research: his verve and passion have always fascinated his students, who were captivated by his love of particle physics. I personally regard this as a great gift in a scientist, and I admire Bellettini for that; I cannot understand those who criticize him for his positive, energetic, inspiring attitude. In the radio interview, Bellettini gave a remarkable answer when asked what he believed was the source of the CDF anomaly:

"Sometimes as I wake up in the morning I think it's new physics, and other times I wake up and think it is just QCD."

Bellettini's interview did not go unnoticed. The perception of the listeners to the NPR interview was that CDF had really found new physics. The phones of many CDF members started ringing that morning, with colleagues from other experiments calling to enquire on what that new find was: a new quark? An even tinier particle? A few of them were caught totally unprepared by the enquiries, as in a large collaboration with many different working groups even a striking result like the anomalous excess at high jet energy could get overlooked.

As mentioned above, several newspapers throughout the USA managed to organize quick articles on the CDF anomaly as they got notified by *Science*. Many of those pieces appeared on February 8. On the *New York Times*, it was Malcolm Browne again, titling "Tiniest Nuclear Building Block May Not Be the Quark." Browne had got a hold of Brenna Flaugher, and quoted her:

"What's different about these results," Dr. Flaugher said, "is that there are many more high-energy transverse jets than theory predicts. The highest energy transverse jets were 120 percent more frequent than theory would explain," she said.

Brenna could not have possibly said "high-energy transverse jets" in place of the correct "high-transverse-energy jets," but the quote was otherwise technically accurate. Another source identified by Browne as "a spokesperson," and which most in CDF would bet was Bellettini, was also quoted as saying "We spent more than a year working on the data to be reasonably sure we weren't making some big error, [...] and now we're confident at least that something unexpected has happened." Again, not a terrible tweaking of facts, yet the mention of the "unexpected" was enough to disturb the most sensitive collaborators. On one thing the writer fully showed he had a clear vision, though, as he concluded his piece with the following forecast: "the CDF paper is expected to undergo intensive scrutiny and criticism in the months ahead."

Curt Suplee on the same day wrote for the *Washington Post* a piece titled "Quark as Basic Particle May Be in Dispute." The quotes contained in his article did not have names attached, and the tone was open-minded:

"It is possible that quarks — the smallest known constituents of protons and neutrons — are not fundamental, indivisible particles, but may be made up of yet smaller entities of unknown nature, said researchers from the Collider Detector at Fermilab" [...] "Until alternative explanations or possible errors have been ruled out, the 450-member team reports in a paper submitted to the nation's leading physics journal, *Physical Review Letters*, any claim about the presence or absence of new physics is not defensible."

In this case, what was most disturbing to CDF collaborators was the rather imprecise description of the findings of the experiment: one read that

"The team turned up nearly 1200 observations during a year of experiments in which colliding particles were deflected or "scattered" in ways that apparently cannot be reconciled with the predictions of current particle-physics theory. But the odd trajectories and energy levels observed in the collisions might make sense if quarks were composed of tiny sub-units that could send matter flying off in unexpected ways."

The above description was frankly substandard, even for a national newspaper more accustomed to deal with politics and football than with particle collisions. It is surprising to note that its author is a renowned and appreciated science writer, who has won during his career several awards for outreach activities. "Odd trajectories" and "matter flying off in unexpected ways" are rather awkward misrepresentations. They offer a good example of the dangers of oversimplification in outreach. Also, the mention of "1200 observations" must have been the result of a complete misunderstanding of the way the experiment collected its data; one which I cannot even trace back to a possible source. 1200 runs? 1200 events above 200 GeV? We will never know, but surely Suplee's article had its share of responsibility for the discontent among the CDF members.

The *Chicago Tribune* also published on February 8 an article on the CDF anomaly, titled "Fermilab's 'Preons': Bad Math Or A Profound Find? As Basic Matter, Quarks Superseded Atoms Now A Flash Of Energy May Shake Up Physics." The author, Ronald Kotulak, wrote it after interviewing Rob Harris, who was careful enough to downplay the practical significance of the CDF find. Harris clearly explained to Kotulak that the theory was probably wrong, as he is quoted to say "If one were a betting person, one would tend to bet that it's more likely to be an error in the calculations than it is to be a new quark substructure." That sentence was the cause of the mention of "bad math" in the title! Because of the *Chicago Tribune* article, Harris got a number of unpleasant comments from colleagues who thought that too much emphasis had been put on new physics, but it really was not his fault. Indeed, to many, the fault was in the way the whole interaction with the press had been managed by the spokespersons. Bellettini, in particular, became the target of the criticism. However, rather than being expressed openly in the appropriate venues, which would have been the Executive Board meeting or the associated mailing list, the criticism remained at the level of corridor rumor, a sort of stomach rumble of the collaboration. As such, it was even worse than a public trial, as Bellettini never got a chance to defend his actions in front of his collaborators.

The only surfacing critique of Bellettini's action was a letter which Avi Yagil posted on February 9 to the CDFNEWS forum, a mailing list where CDF members mainly exchanged news about papers in preparation, meeting times, or newborn babies, but also one which, unlike the Executive Board mailing list, reached all CDF members. In his posting, Avi summarized the result of a discussion of the situation with colleagues who like him resented the way the interaction with the media had been handled. He explained that the spokespersons had the right to speak on behalf of the experiment, but were not supposed to confuse their own opinions with the ones of those they represented.

Avi's summary is believed to have had one notable effect when in 1997 the first term of Bellettini as spokesperson expired: he declared his availability to serve for another term, but he lost the election. It seems as if the media-wary soul of the collaboration did not forgive him for the way newspapers and outreach magazines had covered and over-hyped the jet E_T excess. In retrospect, I do believe Bellettini was guilty: he was guilty of believing that creating attention on particle physics and explaining to laymen the excitement of a possible new CDF discovery was in the interest of the experiment. There lay the mistake: it was in the interest of the progress of science, but against the interest of some of his colleagues.

As for Brenna Flaugher, she was left with a sour taste from the whole story. After the media storm of February 1996, she had to withstand open criticism from her colleagues that had nothing to do with the analysis of the data but a lot to do with how others had been presenting it. She was surprised that people also resented the way she described the hypothesis of compositeness in presenting the results at conferences and in interactions with science reporters. There was nothing new or unusual in testing compositeness with the high-jet-E_T data; theoretical papers had repeated for 20 years that this was one of the most direct ways to test the standard model. She believed that the interest of the press in the science that CDF was doing should have been a positive thing. Eventually, Brenna got more interested in building detectors and let other people fight what were clearly more political battles than physics arguments.

The Anomaly Vanishes

On November 21, 1995 Kuhlmann, together with his colleagues in the CTEQ group, sent to the Cornell Arxiv a first version of their article, which was titled "Large Transverse Momentum Jet Production and the Gluon Distribution Inside the Proton." The article was revised a few months later, and a second version replaced the former one in May 1996. The abstract of the second version read as follows:

"The CDF experiment has reported an excess of high-pt jets compared to previous next-to-leading order QCD expectations. Before attributing this to new physics effects, we investigate whether these high-pt jets can be explained by a modified gluon distribution inside the proton. We find enough flexibility in a global QCD analysis including the CDF inclusive jet data to provide a 25–35% increase in the jet cross sections at the highest pt of the experiment."

In the article one finds a graph which shows how their modified gluon content in the proton PDF produces a model in great agreement with the CDF data. In the revised version, the authors inserted a note which criticized the work of a competing group of PDF fitters, the MRST collaboration. MRST had published a study two months earlier, where they claimed that

"it is impossible to achieve a simultaneous QCD description of both the CDF jet distribution for transverse energies $E_T > 200$ GeV and the deep inelastic structure function data for $x > 0.3$."

In the note added to the second version of their article, the CTEQ group thus comments:

"A recent paper from the GMRS group [20] has stated it is impossible to obtain parton distributions in agreement with the CDF jet data. Their attempt to modify the quark distributions to fit the jet data results in a χ^2 of 20703 for 128 BCDMS data points. In contrast, the two jet-fits presented in this paper give rise to a χ^2 of 173 and 175 respectively for 168 BCDMS data points. It appears the reason that GMRS were not able to find satisfactory solutions like ours is that they did not allow sufficient flexibility in the gluon distribution shape."

The "BCDMS data points" in the above quote are data from a fixed-target experiment which strongly constrained the proton PDF in a kinematic region relevant to the CDF measurement. The numbers mentioned in the above abstract are values of the "chi-squared," a statistical measurement of agreement between data and the model. The larger is the chi-squared of a fit, the worse is the agreement of the data to the used model. So, the CTEQ group fits were much better than those of MRST; the CTEQ parameterization had the flexibility to accommodate the CDF data.

The CTEQ explanation of the anomaly was convincing by itself, but in the summer of 1996 came a very clear experimental proof that no compositeness models were needed to interpret the CDF data. This was the analysis of the dijet angular distribution, which had been carried out by Chao Wei and Rob Harris. The distribution of the angle between high-E_T jet pairs followed the theory prediction bang-on, indicating that the data had the same kinematic features expected for QCD processes. If quarks had a substructure and that were the source of an excess of energetic jets, those jets would betray the qualitatively new scattering mechanism by distributing at different angles from what was observed. Wei and Harris of course could not rule out compositeness models across the board: the existence of a quark substructure at distance scales much smaller than those tested with the energetic Tevatron collisions was (and is) still possible. But their analysis proved that preons were no explanation of the excess of high-E_T jets, effectively ending any speculation.

The jet E_T distribution of Run 1B data was later also analyzed by Anwar Bhatti. By the time he concluded his study, the CTEQ collaboration had produced a new PDF model. While remaining in good match with all available inputs, the new functions included a tweaked-up gluon distribution at high momentum fraction. The CDF data were found to follow the same shape of the 1A result, confirming that the effect was systematic and not due to a statistical fluctuation. Yet, this time the new PDF set explained it well, without any need for a preon hypothesis. Bellettini's preon dreams were thus archived.

Chapter 10

A Personal Interlude

Participating in an experiment that counts half a thousand collaborators means you will never get to know all of them, not even if you spend all your life at the lab. During Run 1, only a minority of the members of the CDF collaboration were based at Fermilab. Most of the others were "irregulars" who visited infrequently. One such occasion was to participate in collaboration meetings. Another was to attend to their data-taking shifts duty, a necessary requirement they had to fulfil in order to be granted authorship rights.

For those like me, who used to spend at Fermilab only a few months every year, it was not easy to get acquainted with everybody else. In particular, it was hard to meet colleagues who did not attend physics meetings, or who spent most of their time closed in their offices. Paolo Giromini was a perfect example of the above. That allowed me and him to live happily for quite a while without a chance to meet, although his office was only a few dozen steps away from mine. When we eventually met, however, we found out we had something to talk about. That was the subject of my Ph.D. thesis, the decay of Z bosons to b-quark jet pairs: the topic stimulated his curiosity for a reason or two.

Just a Check

After the 1973 discovery of weak neutral currents by the Gargamelle experiment at CERN, and the many confirmations of the standard model which followed, the direct observation of the W and Z bosons at the Sp$\bar{\text{p}}$S hadron collider was no surprise. Practically, no particle physicist doubted their existence by then. As stated by Donald Perkins, one of the members of the Gargamelle collaboration,

> "When they [*the weak W and Z bosons*] were found, the discovery was greeted with relief rather than disbelief."

As explained in Chapter 3, measuring Z boson properties in hadron collisions became irrelevant after 1989, as the four LEP experiments at CERN and the SLD experiment at SLAC exploited electron–positron collisions at 91 GeV in the center of mass to produce millions of Z bosons. The CERN and SLAC experiments could then determine with unchallenged precision several quantities which constituted a crucial input to electroweak theory. Regardless, the samples of Z boson decays to electron or muon pairs collected by CDF proved extremely useful during Run 1, as the precise knowledge of the Z mass provided by electron–positron experiments could be used as a calibration point to determine whether the lepton momentum and energy measurements were well tuned or whether instead they required some adjustment.

In contrast to the very clean leptonic decays, Z decays to jet pairs were thought next to impossible to even spot, leave alone exploit, in hadron collisions. Yet, it was something to which I had turned my attention at the start of my Ph.D. By 1996, the signal of all-hadronic decays of top-quark pairs (see Chapter 4) was emerging in the data, thanks to a long-lasting effort by the CDF-Padova group; the analysis was the topic of my undergraduate thesis. The success of that endeavor convinced me that hadronic jets could indeed be exploited for the search of high-mass particles. The field of "jet spectroscopy" was still in its infancy, yet it was clearly the way to go to further the investigations of the high-energy frontier at particle colliders. The question was, what high-mass particles should one go searching for, as a next step after the top discovery? The decay of Higgs bosons to jet pairs came to mind, but that was a lost cause: CDF would only acquire some sensitivity to a Higgs signal 15 years later, after having acquired 100 times more data with Run 2. As for the many exotic new particles hypothesized by supersymmetry or other new physics models, they did not appeal to me: I did not fancy spending years in the search of some non-existent entity and come up with a null result! I wanted to extract a real particle signal from jet events, a mass bump I could take a picture of and hang on the wall.

This proposition left only W and Z bosons on the table. Weak bosons frequently decay to pairs of jets; that signal, although riddled by huge backgrounds from QCD processes, had been observed in 1987 by UA2. But it was not clear whether CDF could repeat the success of the old CERN experiment, despite being a better detector and having collected 20 times more data at a three times higher collision energy. Surprisingly, the higher collision energy of the Tevatron is a disadvantage in this case, because the QCD background of dijet pairs grows much faster than the signal as one increases the collision energy. Furthermore, as I dug more into the matter, I found out that UA2 had used a specially designed trigger which collected unbiased samples of jet pairs; CDF did not have a similar trigger. Yet, CDF had one crucial weapon at its disposal: the capability to identify b-quark-originated jets. The Z boson decays 15% of the time into b-quark jets, which are very rarely found in QCD background events. Maybe by selecting b-tagged jets the signal would show up!

Was it possible to extract Z decays to b-quark jet pairs from Run 1 data? One day I asked that question to Weiming Yao, one of the fathers of the SECVTX b-tagging algorithm. Weiming answered me by raising his arms, as if asking for forgiveness: "The background is too high!" He explained that he and others had, in fact, tried to isolate a $Z \rightarrow b\bar{b}$ signal in the past, but had soon quit because of the impossibility of reducing the QCD contamination.

Rather than deterring me, Yao's assessment became an additional motivation: I would succeed where others had failed! Unfortunately, my enthusiasm did not appear contagious. One piece of evidence in that direction came in the spring of 1997, when I visited Jonathan Lewis in his office to learn some details of muon datasets and to ask for a program he had written to simulate the trigger that collected events with muon pairs (i.e., *dimuon* events). I was trying to determine whether those data would be useful for my search.

Jonathan Lewis, who at the time had a "Wilson fellow" position at Fermilab, was considered a sort of guru of muon triggers: he had spent many years babysitting them. He was in his mid-30s, lean but not tall, with short hair just turning gray. Hiding behind large glasses, his eyes were quite lively, but they would seldom stare at you. The day I met him, Jonathan wore a baseball hat, khaki shorts, tennis shoes, and a checkered shirt, and he sat in the middle of a small office full of tidily organized paraphernalia for his hardware activities, along with a large number of articles and notes distributed around in no apparent order. With my still quite imperfect English, I probably appeared a bit dumb to Jonathan, the more so since I sounded enthusiastic about my analysis project, which did not seem a great idea to him.

"Ah, here you are. So you are the guy who needs a simulation code for dimuon triggers. What do you need it for, by the way?"
"It is for a search of Z boson decays to b-quark jets. You know, it is a signal which has never been found at a hadron collider before. I think I can do it! If I select events from the muon stream and apply a double b-tagging, the signal could well be visible in the...."

He looked bored as he interrupted me:

"I see, I see: you want to find Z bosons. So this is a check, right?"
"Well, hmmm .... What do you mean?" I asked.
"Z bosons are the most studied particles around. There is no measurement we can do here to improve our knowledge of their physics." replied Lewis.
"No, I think it is important! Look, we need to see the Z decay to jet pairs before we hunt down the Higgs, to prove we can do it."
"Sure, that's what I meant — it is a check. But why the dimuon triggers?"
"Well, I am trying to find out which are the most promising datasets to look for the Z. If both b-quarks decay to muons, then you would expect to find a signal in dimuon triggered data." I answered.
"Ok. I'll see what I can do — maybe I can send you a piece of code with some example. Send me an e-mail to remind me, though. I am pretty busy now."

My thoughts on the matter were different back then, but after all, Jonathan was right in his belittling assessment. Finding Z bosons in CDF was, the way a matter-of-fact guy would look at it, nothing more than a check. It surely was not going to be a measurement worth a line in the *Annual Review of Particle Properties*, the summary of all human knowledge on subatomic particles, nor would it result in a heavily cited article. Still, it was a check nobody had managed to carry out before, and this was by itself a motivation. I was determined I would pull it off. Besides, finding the Z → bb signal would be useful to the experiment in at least a couple of ways.

Showing that a resonance could be reconstructed from its decay to two b-quark jets was very important for low-mass Higgs boson searches. The standard model predicts that a light Higgs decays most of the time into a pair of b-quark jets. As the production rate is extremely small, isolating a signal in the jet decay mode requires one to improve as much as possible the mass resolution of jet pairs. By narrowing the Higgs mass peak, one may better see it emerge above the large background from QCD jets. In parallel, the background must be reduced by a kinematic selection exploiting the observable differences between electroweak and strong production processes. The two critical tasks mentioned above may be effectively optimized once one sees a Z → bb signal emerging as a bump over the featureless mass distribution of the QCD background. This is because both the algorithm designed to improve the mass resolution and the kinematic selection can then be tuned on the real signal rather than on imperfect Monte Carlo simulations. In addition, a sizable signal of Z → bb decays could provide CDF with a calibration point to improve the understanding of the jet energy measurement, the dominant source of systematic uncertainty in top mass determinations.

Which Dataset?

Events containing two hadronic jets with an energy corresponding to half the Z mass each, i.e., 45 GeV, are produced at prohibitively high rates in Tevatron collisions: the trigger cannot collect them all. However, only a very small fraction of the jets are originated by bottom quarks. Hence, a trigger which selectively collected b-quark jets would acquire Z → bb events without drowning the data acquisition system. Luckily, bottom quarks are indeed special: besides having a long lifetime, their decay often yields an electron or a muon. CDF was designed to effectively trigger on charged leptons of high-transverse momentum produced by the decay of W bosons: those were the backbone of most top-quark searches. Electrons and muons emitted in bottom-quark decay were harder to spot, because they hid amidst several other particles produced in the b quark hadronization. But triggers collecting low-energy, "inclusive" samples of events containing electrons and muons embedded in hadronic jets did exist. Those events could be collected because they were comparatively rare: only a few of them per second got stored into the "inclusive electron dataset" and the "inclusive muon dataset."

In the spring of 1996, I started my Z search in the electron dataset. At the time I was well aware of the study that a Padova colleague, Alberto Ribon, had been carrying out. He was measuring the properties of B hadrons using muon samples and was having a lot of trouble understanding muons of very low transverse momentum: those data were riddled by large, ill-understood backgrounds. Electrons looked less complicated to study; still, they are tough to find when they are embedded in a hadronic jet, because of the way they are identified. Electrons produce an ionization track in the CTC and then, upon hitting the first layers of lead sheets in the electromagnetic calorimeter, start a shower by the combined processes of bremsstrahlung and pair production, as described in Chapter 2. Because charged hadrons can sometimes produce a similar signal and are produced at a much larger rate, electron candidates must pass very tight selection criteria to make their signature sufficiently reliable. Unfortunately, Monte Carlo simulations underestimated the loss of real electrons of that tight selection and predicted that the dataset contained more signal than it in fact did. That mistakenly led me to think the electron dataset would be the more promising for my search.

Using the inclusive electron sample, my search for Z decays to b-quark pairs relied on some kinematic properties and on the identification of a SECVTX b-tag in both jets. After tuning the selection strategy, I rushed to check whether the invariant mass of the two jets showed any hint of a clustering close to 91 GeV, the mass of the Z boson. Simulations predicted that the Z signal would become clearly visible in the data, yet all I could see in an otherwise smooth spectrum was an inconclusive shoulder of few tens of events, which could have well been due to a statistical fluctuation of the QCD background.

I was depressed by the outcome. But then I started thinking seriously about muons, which I had until then ignored. At first, the inclusive muon sample appeared no more than a means of doubling the statistical power of the search. Yet, muons turned out to be a pleasant surprise. The same analysis cuts that had only produced a faint hint of the Z signal in the electron dataset turned out to select a sample which displayed a large bump in the dijet mass spectrum!

For a while, I was quite suspicious: I thought electrons and muons should give a similarly sized signal. I had not yet been able to work out an accurate estimate of the signal efficiency for the muon selection, but the two datasets appeared to yield conflicting results. How was I going to defend that twice larger muon signal in front of the collaboration? It was Barry Wicklund who solved the enigma for me, as I bumped into him at the entrance of the B0 building one morning. Coming down from the third floor with a styrofoam cup of coffee in my hand, I turned the round knob of the door leading out. While I turned and pushed from the inside, Barry turned and pulled from the outside. The door opened considerably more quickly than I predicted, and the coffee washed out on my pants and on the floor.

"What the — Whoops!"

"Oh, sorry! Everything okay?" asked Barry.

"Well, I guess I'll have to go get another cup, but yes, I'm fine. The coffee was not too hot." I replied.

Barry politely took a chance to inquire: "How is your Z search going, by the way?"

"Oh, quite well, thank you for asking! … Although there's something I can't quite figure out yet." I explained.

"What is it?"

"Well, it's a bit premature to say, but it seems there is a large signal in the inclusive muon sample, when I do the same data selection that isolated a small bump in the electron sample — you know, the stuff I showed at the last electroweak meeting; I think you were there." was my reply.

"Yes, I remember. And where's the problem?"

"I mean, shouldn't I be getting the same kind of signal with muons? I think the integrated luminosity of the muon sample is actually a bit smaller than that of the electron sample…."

"Yeah, but electrons have a much lower efficiency to be tagged inside jets," bolted out Barry.

"What do you mean?" I held my breath as he answered.

Barry explained: "Of course, to be classified as a good electron, the candidate must have an energy deposit in the calorimeter with a large electromagnetic fraction, which is difficult if the electron is inside a jet, as nearby hadrons produce energy in the hadronic calorimeter."

The electromagnetic fraction is the ratio between the energy measured in the inner electromagnetic section of the calorimeter and the total measured energy, which includes that recorded by the outer hadronic section. Electrons develop cascades of secondaries much earlier than do hadrons, so their energy deposition is usually contained within the electromagnetic calorimeter. That is one of the main handles to spot them when they are contained within a jet: one follows a track to an electromagnetic deposit and verifies that there is not much additional energy deeper down.

Barry continued:

"No similar cut is applied to muon candidates, of course. So their efficiency should be about twice as high."

"So you mean to say that one could expect to see more Z's in the muon data?" I asked.

"Yes, it would be strange if it weren't so." was Barry's reply.

Barry Wicklund had made my day. I rushed to my office and quickly forgot about my badly stained pants: the signal was probably real, and I could now take it seriously.

The numerical significance of the excess I had spotted in the muon dataset was larger than three standard deviations: a "solid evidence" of the $Z \rightarrow b\bar{b}$ signal. I was happy to have explained away the discrepancy between electron and muon samples, but the signal in the latter dataset still appeared numerically too large. My selection extracted 588 events with two b-tagged jets and kinematics as similar to Z decay as was reasonable to ask, while my background estimate predicted I should find 497: there was thus an excess of 91 events over the predicted background, which could well be attributed to the Z signal. On the other hand, the simulation only predicted I would see 40 events or so from Z boson decays in the data. The excess carried a large uncertainty associated with it, as it was the result of subtracting the not very precise background estimate. Still, reconciling 40 or so events expected from the Z signal with the observed excess of 91 was difficult. Barry had clarified to me that the simulation of muons was trustable, so maybe the large excess was a spurious one, or at least in part not due to Z decays? Luckily, I was soon offered a solution to that riddle as well.

"Would You Laugh Less Loudly, Please?"

The office next to mine in the CDF trailer complex was occupied during those years by Fotios Ptohos, who had been a Harvard University graduate student for a while now. That was not such a peculiarity: it was generally understood that in Harvard, Ph.D. studies would routinely last 10 years or even longer. That fact was confirmed by the joke that circulated in the Harvard University High-Energy Physics Laboratory, according to which the basement of the Physics building was full of G-10: no, not the material used to build electronic boards. Rather, 10th-year graduate students!

Fotios was tall and lean, and he had long, light-brown hair which was thinning above the forehead. He wore round glasses and an irregular blondish beard. He was quick with a joke, and I could often hear loud bursts of laughter through the plywood wall that divided our offices, any time Giromini paid him a visit. I had had few chances to interact with Fotios in the past: social life was close to non-existent for the doomed graduate students of CDF.

One evening in the spring of 1998 their voices appeared louder than usual. They were stationing in front of my door, and they sounded like Bavarians at the Oktoberfest. I had been spending the afternoon in the attempt to find a bug in a program which in my intentions was to produce a fancy histogram, but was instead returning a core dump, crashing without appeal. I was frustrated and, with no progress to show for two hours of investigations, I decided to ascribe it to the noise. I thus got out of my door and stared with a frowning face at the duo of jokers. This was the first time I met Paolo Giromini.

It took Paolo and Fotios less than one minute to have me join their laughter; they were commenting on the funny habits of a colleague. For Paolo, it was business as usual: he would joke about anything and anybody; be it the Pope or Sheldon Glashow, nobody was immune from his sarcasm. But soon, they restarted to discuss physics. The issue was the infamous

SECVTX scale factor. The scale factor was defined as the ratio between the SECVTX efficiency to tag a b-quark jet in data and in the Monte Carlo simulation. It was a crucial input to top-quark cross-section measurements. There, imprecisions in the simulation of b-quark jets could lead to large uncertainties on the efficiency of selecting b-tagged signal events, which propagated into the cross-section estimate. Paolo and Fotios explained that they had recomputed the scale factor using a much larger Monte Carlo dataset than the one officially used for that purpose. The generation of that simulated sample had cost Fotios months of unforgiving, repetitive work. With millions of simulated bottom–antibottom production events, the statistical uncertainty on their estimate of the scale factor was now much reduced, and the departure from unity of the latter was unquestionable: that proved that the simulation underestimated the efficiency. The factor was close to 1.25.

In my analysis, I had never considered the need for a scale factor to correct the simulation. Hearing about the matter, I soon realized that my attendance to the Top group meetings had been too scarce in the recent past: the SECVTX scale factor had a large impact on my own analysis!

"Wait, let me see if I understand this," I told Paolo, "you mean to say that despite all the tunings and tweakings of our detector simulation, if one looks for a signal yielding b-quark jets in a simulation done with the Herwig Monte Carlo generator, one is going to see 25% less of it than what one would see in the data, other things being equal?"

"Well, 20% less, as 1 divided by 1.25 is 0.8. The factor applies to the data, or its inverse to the simulation. And Herwig or Pythia make no real difference."

Herwig and *Pythia* were the two most widely used programs that simulated hard proton–antiproton collisions. I continued my inquiry:

"Ok, ok. So if my simulation predicts I should see 40 Z boson decays into b-quark jets, I should rather expect to see 50 in real data, correct?" I inquired.

"No, that depends. It would happen only if your selection required one SECVTX b-tagged jet in the event."

"But my selection requires *both* jets to be b-tagged with SECVTX! Wait …Am I underestimating my signal by 1.25 *squared* then?"

"Well, yes, correct." Paolo replied. "There is no correlation between the scale factor of two b-tags in an event. We see no indication that there be any. So yes, you should square it if you require two b-tags."

"But then my Monte Carlo prediction becomes ... 40 times 1.25 squared ... Hmm 1.2 squared is 1.44, plus 0.12...1.56 times 40...63 events! Sixty-three! It agrees with the data!!!" I enthusiastically replied.

"Wait, what's the excitement? Are you seeing Z decays in the data?"

"Sure, I have isolated a sample where I have an excess of 91 events, give or take 20. Now 91 ± 20 is really off from 40, but it is less than 1.5 standard deviations away from 63, the number I compute if I take into account your scale factor! Thank you!"

"Ah! Bravo guaglione." was Paolo's remark.

The conversation had proceeded in English because Fotios did not understand Italian well — i.e., he knew the meaning of some of the more colorful expressions used by his boss, and little else. His *Bravo guaglione* (good boy), spelt with a southern accent, was a joking way to mimic a mafia boss appreciative of the work of his disciple. Mafia is no laughing matter, but again, Giromini liked to joke with anything. He continued:

"If what you say is true, that would be a nice check of the value we find for the scale factor. Please explain how you find the signal. Do you have a plot?"

I did have a plot; I could have submerged him with plots. The rest of the afternoon was not enough to show him all my work in detail, so our spare time in the following days was spent discussing my analysis. Paolo would criticize without mercy each and every statement I made, mocking me and calling me names when I failed to prove my point: that was his way to do physics. But I seemed to have a defensible signal in my hands. After convincing Giromini that my Z signal was sound, I felt my result could withstand a collaboration approval.

My acquaintance with Giromini turned out to be quite timely, as very soon, as I struggled to finish my Ph.D. thesis, I was drawn into the middle of what would become the biggest and longest-standing controversy in the history of CDF, fueled by an analysis of the Frascati group: the Superjet affair.

Chapter 11

The Superjets Affair

In 1995 a group of physicists from Pisa and Frascati led by Paolo Giromini finally completed a measurement which had kept them busy for several years: the determination of the total cross section of 1.8 TeV proton–antiproton collisions, a number which is a fundamental input to any determination of the luminosity of a data sample. In the early 1980s, the Italians had built a set of detectors specifically designed to track small-angle particles emitted in the very forward direction. Those detectors, called "roman pots," provide an accurate measurement of the rate of interactions that do not break up the protons. A detailed study of the resulting data allowed a threefold decrease in the systematic uncertainties on the total cross section. This translated directly into a reduction from 15 to 5% of the uncertainty in the integrated luminosity of all CDF datasets. It was an important contribution to the scientific output of the experiment, as the precision of the cross section of all rare phenomena measured by the experiment improved by the same factor.

Once freed from that grievous task, Giromini could turn his attention to more interesting physics investigations. The measurement of the top-pair production cross section was a natural choice, a logical consequence of his previous work: the top quark was a new particle and its production rate needed to be determined as precisely as possible; in addition, of course, top quarks constituted a new, exciting research topic. Yet, as soon

227

as he got his hands on the Run 1 dataset of W+jets events, Giromini found even more reasons to be interested: a hint of what could be interpreted as a Higgs boson signal appeared in the data!

Higgs Alert

Until the second half of the 1990s, when the LEP collider was upgraded to produce electron–positron collisions at a higher center-of-mass energy than previously achieved, the Higgs boson (see Chapter 1) was not the main focus of experiments exploring the high-energy frontier. The reason is that the expected cross section of that particle was prohibitively small for the comparatively low energy and luminosity provided by the facilities then available. One could still look for anomalously high-rate production of final states possessing the characteristics of a Higgs boson decay, but those searches had a limited appeal.

A study of the future chances of a Higgs boson discovery with CDF in Run 2 was provided by Steve Kuhlmann, who in 1995 determined how the Higgs signal could be evidenced by searching for its production in association with a W boson. The Higgs boson can be produced through many different reactions in proton–antiproton collisions. The direct production of a Higgs boson without other particles is the most frequent of them, but it is generally quite hard to spot it because most of the time the Higgs decay yields just a pair of b-quark jets. The resulting rather feature-less dijet events cannot be collected effectively by the trigger system due to the extremely high rate of competing QCD backgrounds. The associated production of a Higgs particle with a vector boson (a W or a Z), while occurring with a 10 times smaller rate than direct Higgs production, offers a much better chance of a sensitive search at the Tevatron. The vector boson decays to energetic lepton pairs provide effective triggering signatures. Kuhlmann had considered how the Higgs decay to a pair of b-quark jets could be reconstructed and had started to study in detail how custom-made corrections to the measured jet energy could improve the dijet mass resolution, boosting the sensitivity. His work showed that a discovery was possible, although it required large integrated luminosity — a long Run 2. Kuhlmann's study turned out to be crucial for the future of CDF as soon as it became clear that the case needed to be made for an extended high-luminosity run of the Tevatron after the end of Run 1 and the foreseen upgrade of the accelerator.

After the 1993 cancelation of the SSC project, the Fermilab director John Peoples realized that there was no prospect of building a big machine in the USA that could compete with the Large Hadron Collider (LHC), the proton–proton collider planned to be built at CERN. He also expected that the LHC, whose design and development studies progressed unhindered, would overwhelm the Tevatron after a year of commissioning, thanks to its seven times higher energy and the foreseen order-of-magnitude higher luminosity. As long as the Tevatron was alone at the energy frontier, it would have a high priority at Fermilab. Peoples had made construction of the Main Injector, a new accelerator providing intense beams for the Tevatron, the highest priority of the laboratory in 1990 because it provided a path to high-luminosity studies of the top and whatever else could be reached with 2 TeV collisions. After 1995, he made the measurements of neutrino properties with the NUMI and MINOS experiments the next highest priority of the lab, as he understood that neutrino physics was a promising door to new physics. A powerful neutrino facility would provide Fermilab with an affordable future, along with the new and promising developments of astro-particle studies. In summary, Peoples strongly believed that Fermilab needed a plan for its future that did not depend on the Tevatron in the long term. The laboratory would have at least until 2006 to operate the Tevatron and collect large integrated luminosity before the LHC start-up turned its competitor into a second-rate machine. The lab director rejected the plan of directing all the efforts of the laboratory into Run 2 of the Tevatron, because he felt this could be a threat to the laboratory's future. The CDF and DZERO upgrades for Run 2 were a very visible and important project, but they did not integrate well with other research efforts that the lab had the need to pursuit.

While Peoples' plans appeared sensible, many CDF members were worried: the Tevatron was the highest energy collider in the world and it would remain such for another decade. It had to deliver as much data as possible to the experiments before one could decide its future. However, changing the director's plans entailed making a very strong case for an extended running; the experimentalists had to join forces, as an effort from only one of the two experiments had few chances to succeed. Dante

Amidei was among those who started to think deep about the issue. On an autumn day in 1994, he chanced to meet his DZERO colleague Chip Brock at the Chicago O'Hare airport. They soon found out that they were both worried for the future of hadron collider physics in the USA, and they ended up spending the time before their flights discussing how to put together an effort to increase the scope of Run 2, and how to make a strong physics case for it.

At first, it looked as if the top quark should be the main motivation for an extended study: that particle was on the verge of being discovered, and the discovery would bring media attention and leverage to the experiments. The top quark was very heavy, and this fact alone could be considered a hint that new physics hid in its phenomenology. Maybe its decays would differ from predictions, or processes involving the newfound particle could become a tag of new physics. Supersymmetry, for instance, could in some scenarios order its squarks into an "inverse mass hierarchy," such that the super-partner of the top quark would be the lightest one. Given the natural expectation that quarks and squarks of the same generation would be coupled more tightly to each other than across different generations, this made the study of the top quark a potential way to detect supersymmetry.

Soon a grass-roots movement took shape in CDF and DZERO, and a project called *TeV 2000* was started by a meeting held at Michigan University on October 21, 1994. That event was attended by over 100 Fermilab physicists. The case for a long Run 2 seemed hard to defend with top physics alone; accordingly, studies of supersymmetry, electroweak physics, and exotic physics were started within independent working groups. And then there was Kuhlmann's study on the Higgs: as soon as Amidei saw it, he understood that it was perfect for the scope of the report that was being produced. Kuhlmann had estimated the chances of a Higgs discovery as a function of Higgs mass and available integrated luminosity. In CDF note 3342, titled "Will we find the Higgs in Run 2?," he summarized his study by noting that a 100–130-GeV Higgs would be hard to discover with only 5–10 inverse femtobarns of data, but 25 inverse femtobarns could be sufficient.

Today, Kuhlmann's original 1995 assessment appears amazingly accurate, given that the now discovered 125-GeV Higgs produced a little less than a 3-sigma effect in the 10 inverse femtobarns of data collected by the Tevatron until 2011. His was indeed a great study, and it became the pivot point to steer the laboratory toward a stronger commitment to Run 2. During all reviews of the TeV 2000 proposal of extending the scope of Run 2, the audience would immediately resonate as the Fermilab physicists mentioned that there was a luminosity threshold above which the experiments would have a clean shot at discovering the Higgs boson before CERN.

While the TeV 2000 studies were edited by Amidei and Brock in a public document, Paolo Giromini became acquainted with the analysis tools and the datasets in use by the Top group. Initially Paolo spent some time with Weiming Yao, who taught him all the nuisances of SECVTX and its correct use. But very soon, the two researchers started to work independently, both focusing on the subset of events in the "two-jet bin" of the W+jets data sample which everybody used for top-quark studies. In that subset of events, an excess over the background estimate had become apparent when one selected data containing a SECVTX b-tag in one of the two jets. By itself, the excess was not numerically very significant. However, both Paolo and Weiming had realized that the invariant mass of the two jets in those events appeared to cluster around 110 GeV, quite unlike what was expected if those events belonged to W+jets or top production. As Paolo used to say, the W+jets background was "as flat as Illinois."

The hypothesis that the W+jets data contained a small signal of a new particle decaying to b-quark jet pairs was exciting. The excess of events in the two-jet bin, if interpreted as due to an extra production process, implied a cross section much larger than what the standard model predicted for associated WH production. If it was a true signal, this would not only be the discovery of a Higgs-like particle, but also a clear sign of the standard model finally breaking up!

While Weiming was accustomed to working alone, and he knew all the tricks of b-tagging, Paolo felt disadvantaged. After the effort with the total cross section measurement, the members of his group had started different analyses; although they were willing to help, they could not guarantee the focused effort that the urgent study required. Paolo needed more manpower, so he asked for the help of his colleagues from Pisa University. CDF-Pisa was a big group, but it was split into a few wholly different analysis activities: top physics, B physics, QCD. Bellettini himself had failed five years before to steer the group's analysis efforts toward the top-quark quest. It proved no easier for Paolo to persuade the Pisa analysts that the W+jets data contained a signal worth pursuing with undivided focus. However, he got lucky with the Pisa researchers who were at the time involved in the construction of the Silicon Vertex Trigger (SVT). SVT was an innovative hardware system designed to perform a superfast but still quite precise online reconstruction of charged particle tracks at Level 2. In Run 2, the system would revolutionize the B physics program of CDF by allowing the collection of events based on the presence of tracks with a significant impact parameter. As SVT was a longer-term project than the various ongoing analyses, this gave comparatively more freedom to its participants. In the end, Stefano Belforte and Giovanni Punzi agreed to help Paolo. In exchange, the latter promised to help them with the commissioning of the SVT in the future.

CDF Note Wars

A tight competition ensued, as both Paolo and Weiming wanted to be the first to produce a thorough interpretation of the data and obtain a conclusive answer on the nature of the potential new signal. Weiming wrote up some interlocutory results in CDF note 3430, which was uploaded in the archive on December 2, 1995. On the following day, the Frascati group produced CDF note 3431, which was more daring as it openly spoke of a preliminary evidence of a bottom–antibottom quark state already in the title. The same pattern occurred a month later, when both groups were able to document more detailed and advanced studies: Weiming produced an updated version of his note on January 12, 1996, and the Frascati group posted five days later their own update as an independent document, which got archived as CDF note 3485.

What did those documents contain? Both Weiming and Paolo focused on the numerical excess of 2-jet events with b-tags, and then both considered the mass distribution of the candidates to show that it was not like what one expected for backgrounds. But the similarities ended there. Weiming reported on a detailed study of the different possible hypotheses for the origin of the observed excess in terms of standard model processes. The Frascati researchers instead aimed at showing their good understanding of the top production contribution to W+jets data, as a preliminary step. They also considered Z+jets data in their search, which was a reasonable thing to do as the Higgs boson could appear together with a Z boson through the same associated production mechanism that yielded WH events. Finally, they included as "double b-tagged" events in their signal search ones where one single jet contained both a SECVTX and an SLT tag but the other jet was not tagged by either algorithm.

Both studies were highly suggestive of the presence of a new signal in the data. The Frascati note contained a one-liner as concluding chapter, quoting from a recent movie: "'It looks too good to be true' — Ray 'Bones' Barboni, *Get Shorty* (1995)." It is also remarkable that despite the competition, both documents recognized the contribution of the other group: Weiming Yao wrote in his acknowledgments section "[...] Special thanks to P. Giromini [...] for [...] comments on the early draft." The Italians in turn wrote "We are indebted to [...] Weiming Yao for teaching us the use of the b taggers [...]."

The CDF notes of Weiming and the Frascati group were immediately recognized as highly sensitive material by their authors as well as by the managers of the experiment. The documents were even equipped on their front pages with explicit disclaimers typeset in boldface to forewarn the internal nature of the information. Yet this did not deter some of their collaborators from openly sharing the exciting news with colleagues around the world. For example, a copy of Giromini's confidential document was found one evening by a CDF collaborator at CERN, on the desk of a respected member of a LEP experiment. Apparently, the more an internal result was surprising and potentially ground-breaking, the more likely it was that a leak of the sensitive information would take place!

The excitement about the possible anomalous Higgs decay signal in W+2 jets data reached its climax in January and February 1996, as physicists from the University of Chicago decided to triple-check the studies of Weiming and Paolo. Mel Shochet and their collaborators collected on January 29 their observations on the WH search in CDF note 3502. In that document, they described their repetition of the Frascati analysis; they could find only a subset of the event candidates listed in CDF note 3431. Alas, this was quite typical: the intricacies of data reprocessing with different software versions, ever-changing calibration constants, and other slightly arbitrary inputs made it extremely hard for any two independent analyses of the same dataset to come out with identical results. But while the Chicago physicists in the end did not find results in big conflict to the Frascati analysis, they objected to the procedure of *jet merging* that the Italians were using. That was one of Paolo's ideas, based on his observation that in WH simulations the Higgs boson decay would sometimes produce three jets instead of two, as one of the b-quarks radiated a hard gluon before hadronizing. Paolo had decided to look at jets identified with a narrow $R = 0.4$ clustering radius (the standard one in top-quark analyses), and then "merge" pairs of jets if they were close in angle. The procedure was not in principle a bad idea: its rationale was similar to the one motivating Ray Culbertson's choice of an $R = 1.0$ radius in his own search for bottom–antibottom resonances associated with a photon (see Chapter 8). However, Chicago showed that the Frascati method was ill-founded in the case of the double-tagged events which made up the tentative Higgs signal, since many of the merged jets could not be interpreted as due to soft QCD radiation effects. The other criticism of Chicago concerned the inclusion of events with a SECVTX and an SLT b-tag in the same jet in the definition of "double b-tags," a description that had until then been reserved for events with two different jets both b-tagged.

On February 7 the Frascati group responded to the Chicago document with CDF note 3521. In it, the Italians argued that the different event selection results of Chicago were understood as due to different calibration constants and slight differences in the starting datasets, plus the use of different software releases for the event reconstruction: the discrepancy did not look like a big deal. The Chicago collaborators replied just two days later with a short document, CDF note 3524. The note cleared some of the confusion on the reasons why the two groups were finding different event candidates, and explained better what was the culprit of Chicago's objection to the Frascati analysis: the use of events which contained a SECVTX and an SLT tag on the same jet. To them, "double tagged" could only mean that *both* jets contained a b-tag, so they did not understand the choice of accepting in that category events with just one b-tagged jet, albeit doubly so. Further, they found it unreasonable to exclude from the double-tag category events with a SECVTX tag in one jet and an SLT tag in the other, which should be *a priori* a less background-ridden sample. The rationale is that one wants to select pairs of jets that both originate from b-quark hadronization, as expected from Higgs decay products, while cutting background events that produce "fake b-tags," i.e., jets originated by light quarks or gluons that get wrongly b-tagged. The Chicago note also had an appendix containing the slightly surreal texts of the email exchanges between the two groups, which tried unsuccessfully to clarify the confusion on the mismatching event candidates. Curiously, the document concluded by noting that the best way to resolve the issues was to go back to exchanging mail messages rather than CDF notes.

By reading those documents, one detects the genuine attempts of the two groups at converging to an agreed-upon definition of the sample of event candidates used to assess the Higgs production hypothesis. The rivalry which existed between some of the Chicago members and Paolo Giromini is lost between the lines, but it was a clear presence during the meetings where the Higgs-lookalike signal was discussed. The argument resurfaced with inverted roles when Ray Culbertson's anomalous dijet peak in photon events (see Chapter 8) was targeted by Paolo and his collaborators. They studied Ray's tentative signal and critically assessed its nature, producing on July 3 their results in CDF note 3743. In that document there is no direct attempt at reproducing exactly the Chicago analysis. Rather, the note showed that the application of what Frascati considered a sound and well-motivated data selection did not produce any excess of b-tagged events over background predictions in the final state studied by Culbertson. Nor was the invariant mass at odds with the expected background shape, once jets got corrected with a custom-made calibration routine that the Frascati group had designed to specifically improve the energy resolution of b-quark-originated jets. The "Conclusions" section of the note is again a sarcastic one-liner, in pure Giromini style:

> "Squeaky clean photons don't carry new physics. New physics hides in the murk."

That statement sounds like a letter of intent, if read with hindsight: in fact, Giromini was going to spend a long time handling the murk of soft-lepton datasets, as I will explain in Chapter 12.

As for the tentative Higgs signal, the anomaly slowly went out of fashion and got forgotten. Perhaps, the Chicago study did end up convincing Paolo that the bump he had isolated in his "custom-made" double-b-tagged selection showed some pathology: besides the ill-justified merging of jets which appeared as genuine and independent objects, the event candidates included a nagging number of jets hitting *calorimeter cracks*. These are narrow angular regions of the calorimeters that are devoid of sensitive elements, as they are crammed with power cables and other service hardware. The energy resolution of those jets had to be bad, so the events which contained them were not expected to pile up as tightly as they did in the mass distribution, even assuming that they were due to the decay of a narrow resonance.

Once they understood that there was no objective support for a claim of a new particle in the data, Giromini and Yao re-joined their forces, producing together a document which summarized their results. The anomalous Higgs signal was thus archived. Yet, the excess of events in the 2-jet bin was there to stay: some of those events would in fact become the basis of new, even steeper, claims.

The Scale Factor Controversy

The collaboration had found a general consensus on the top-pair production cross-section measurement, which got published in the "top observation" PRL article of 1995. Yet, things were bound to change as the full Run 1B data became available for analysis. Fotios Ptochos started to work at that topic for his Ph.D. thesis under the advisory of Melissa Franklin, teaming up with the group led by Paolo Giromini. Unfortunately, his effort was on a collision course with the one of a group of physicists from Rochester, Urbana-Champaign, and Fermilab. The Americans had been among the main authors of the top discovery analyses in 1995, and they now wanted to produce an updated result. As the analysis techniques had been studied in detail and the major issues had been understood and addressed, a more precise determination would cost less effort and would be well worth doing. It would certainly become a world's best measurement, and the article describing it would collect many citations.

There were additional reasons to make the top cross section measurement a very appealing one. In general, the top quark inspires awe: it is the heaviest "point-like" particle known to mankind, a *lusus naturae* of sorts; an anomaly worth knowing everything about. Hence, studying the top quark was a priority at the Tevatron, and being a "top quark expert" was a coveted goal for a researcher. Many collaborators chose to study how to improve the mass measurement by using more advanced fitting methods or by widening the acceptance of data selections. Others set out to study ways to determine the other still not well-measured physical properties of top: charge, spin, and production modes. And then there were searches for non-standard decays of the new quark, and searches for new physics connected with top-quark production.

The diversified effort of CDF in top physics matched well with the scientific goals of the experiment. Having discovered a new form of matter it was mandatory to study it in as much detail as possible, especially since the top quark was going to be an exclusive target of the Tevatron experiments for at least another decade. Yet, one could see more mundane reasons for the interest in top physics of the collaborators. Taking an active part in the publication of a new top physics result had non-negligible implications. It was a visible contribution which often paved the way to more important charges. If we examine the list of spokespersons who alternated at the leadership of CDF since the top discovery, we find three out of ten of them — Young-Kee Kim, Rob Roser, and Jacobo Konigsberg — who were previously conveners of the Top group, and contributed to measurements of the top-quark mass (Kim) and cross section (Roser, Konigsberg): top physics paid top prizes. The advantages were also significant outside the collaboration, as a direct contribution in a top-quark physics measurement was worth highlighting in one's Curriculum Vitae.

Given the conflicting interests and the personalities involved, a clash was inevitable between the Frascati group and the Americans. And indeed after the top discovery, which had made everybody happy and had strengthened the mutual trust of the collaborators, the Top group soon returned to be an unfriendly place. Fotios Ptochos, who regularly presented there the progress of the Frascati studies, started to experience the same treatment reserved to Grassmann in the past. The endless arguments around the same issues and the lack of convergence of their top cross section analysis were the main causes of the autarchic behavior that the group led by Giromini would soon start to display.

The central issue around which the war was being fought was the determination of the SECVTX scale factor, the ratio of b-tag efficiencies measured in real and simulated data. The American group based its estimates of that number on an official, well-understood, and agreed-upon sample of Monte Carlo events. They did not find any significant departure from unity of the number: this meant that the simulation of top production could be trusted to yield the correct rate of top events in b-tagged data, without the need of a tweak. Instead, the Frascati group argued that their analysis could measure the scale factor more precisely and returned a value significantly larger than 1.0.

The SECVTX scale factor can be estimated using large control samples of data enriched in b-quark jets: the inclusive electron and inclusive muon datasets. Inclusive lepton events are selected by the trigger system because they contain an electron or a muon with transverse momentum larger than 8 GeV. Those triggering particles are typically originated in the decay of a bottom quark and are contained within the radius of the resulting hadronic jet (called *lepton-side jet*), which usually recoils against a second jet (the *away-side jet*). One may picture those events as due to the production of a pair of bottom–antibottom quarks kicked out of the interaction point in opposite directions; these create a two-jet topology. The application of the SECVTX algorithm to the two jets in real data, once the fraction of real b-quark jets is known, determines the b-tagging efficiency —that is, how likely it is to find a secondary vertex in a real b-quark jet. The same procedure carried out on the simulated sample of b-quark jets allows one to estimate the b-tagging efficiency on simulated b-jets. The ratio of the two numbers is the scale factor: a number equal to 1 if the simulation is perfect. Note that if the fraction of lepton-side jets and away-side jets originated from b-quarks were known, the measurement of the scale factor would be rather simple. But the fraction is unknown and must be estimated. The recipe for doing that is complex enough that the matter lends itself to endless debates.

The battle over the scale factor raged for several months, but the outcome was never really in discussion: the Americans had a much larger support within the Top group and in the whole collaboration. By the end of 1997, a new measurement of the top-pair production cross section, which averaged together the results of all decay channels, was submitted for publication. The measured value was now 7.6 picobarns, with a 20% uncertainty. It was higher than theory predictions, but not in disagreement with them. It was also perfectly in line with the former CDF measurement. Yet, due to the assumption that the SECVTX scale factor was 1.0, it was a significantly biased estimate, as would much later be proven.

The Frascati group had lost the battle, but they continued to work on their analysis with obstinacy. During 1997 and 1998, Giromini and collaborators studied in detail the reasons why the scale factor they were measuring was larger than unity. This could only be due to shortcomings of the simulation: for example, a wrong modeling of the behavior of b-quarks. To study that hypothesis, they checked the database used by Monte Carlo programs to model the decay of B hadrons and found out that it contained quite outdated and partly incorrect information. The Italians fixed the database by including the new results on B hadron phenomenology obtained by CLEO, an experiment studying low-energy electron–positron collisions at the Cornell laboratory. Those results produced an increase in the chance of finding tracks with large impact parameter in the decay of bottom mesons and baryons. The increased rate of those tracks produced an increased efficiency of finding secondary vertices with SECVTX, explaining part of the departure from unity of the scale factor. The remaining difference could be proved to stem from an insufficiently precise description of the geometry of the silicon detector in the detector simulation, and by the effect of nuclear interactions of hadrons in the silicon layers, which had not been modeled with enough care.

The Frascati study had conclusively solved a long-standing mystery. Yet, the opposition to those results continued, despite the fact that the article produced by the Rochester–Illinois–Fermilab group was now already in print. The reason of this was twofold. First of all, the authors of the previous measurement did not like that a new CDF publication based on the same data would revise their result; after the early 1994 measurement in the "evidence" paper, this would be the second time in a row, and in the same direction (see Chapter 7)! But there was also a deeper, more disturbing reason. The heart of the matter, and the reason of the effort of the Frascati group as well as the opposition to it, was really the true numerical value of the cross section. Let me explain why.

The Italians were finding a scale factor equal to 1.25, now with an uncertainty of just seven hundredths. This difference from unity made the efficiency of collecting b-tagged top events larger by a full 25% than previously estimated. This meant that the cross section measurement would need to be adjusted *down* by one-fifth. A smaller top cross section resulted in a better agreement with theoretical predictions: the measurement ended up sitting at 5.1 picobarns, in great match with the best theoretical estimate then available, 4.9 picobarns. However, like a cover too short being pulled all the way in one direction, the new result got seriously at odds — by more than two standard deviations — with the 9.2 picobarn cross section which one could alternatively measure using the SLT algorithm alone as a b-quark tagger. Also, the smaller SECVTX top cross-section would make the number of events with SECVTX b-tags found in the W+2 jet bin stick out more significantly over the sum of estimated background plus top contribution. Those were the events on which the anomalous Higgs hypothesis discussed earlier in this chapter had been based.

The two anomalies just mentioned, taken together, were an invitation for creative explanations. Giromini reasoned that if some exotic process yielding a W+jets signature were only marginally selected by SECVTX, but contained instead a sizable number of SLT leptons inside jets, it would produce exactly the observed pattern: an excess in the W+2 jet bin, and a high SLT-based measurement of the top cross section. On the other side of the barricade, the Americans did not want to hear about fantasy speculations: to them, the discrepancies were irrelevant and any claim of something odd in the top-quark datasets was anathema, as it would cast doubt on the whole scientific output of CDF in the top sector.

Superjets!

The apparent discrepancy with expectations of the data contained in the W+2 jet bin had been a motivation for two years of study and a dozen of internal CDF notes, but it was the SLT excess of top events which had now become the real puzzle for Paolo Giromini. He suspected that there could be a weird class of events mixed in with genuine top decays in W+jets data, even weirder than the originally hypothesized anomalous Higgs. His first attempt at explaining the discrepancy between the top cross section he measured with SECVTX and SLT involved the hypothesis that the jets in the excess events were originated from charm quarks of high energy. Charm-quark jets have a much lower chance of producing a secondary vertex: the lifetime of charmed hadrons is less than a third of that of bottom hadrons, and their mass is smaller. Those characteristics conspire to yield jets with fewer charged tracks, with less significant impact parameters. In fact, the SECVTX algorithm only reconstructs secondary vertices in about 10% of the charm-originated jets, as compared to the 40% of bottom-quark ones. As for SLT, charm jets produce leptons at about half the rate of bottom quark jets: again the tagging efficiency is lower for charm than bottom, but twice less so. The net result is that a hypothetical charm contamination would produce a larger relative effect in SLT b-tagged samples than in SECVTX b-tagged ones. A hypothesis could thus be cooked up to explain the observed difference in the two measured cross sections: some new physics process yielding charm jets and a leptonic W decay contaminated the W+jets sample. The contamination, mistakenly assumed to be part of the top signal, could be causing an overestimate of the SLT-based cross section of top production, leaving the SECVTX-based one almost unaffected.

It was because of those early claims by the Frascati group, the suggestive documentation produced, and the controversial nature of the slight numerical discrepancies that many members of the Top group grew diffident of the whole package of analyses that Giromini was pushing forward. That was an understandable reaction. The charm hypothesis was objectively on shaky ground, as there was hardly any credible new physics model that could produce a similar signature. In fact, that model was soon discarded in favor of something even more exotic. The analysis of the Higgs-lookalike signal in the W+2 jets bin and the long-range study aimed at determining the top-production cross section eventually brought Giromini to examine in detail the properties of W+jets events which contained jets tagged by SECVTX, jets tagged by SLT, jets not tagged by either algorithm, and jets tagged by both. By determining the expected background in each category, bin by bin in the histogram of jet multiplicities, Giromini could trace exactly to the source of the observed numerical anomalies. The source was a set of events containing jets tagged by both algorithms. Then a *eureka* moment arrived.

The discrepancies with predictions came from an excess of W+2 and W+3 jet events where the same jet was tagged by *both* SECVTX and SLT. There were eight 2-jet events and five 3-jet events of that kind: 13 events all in all. Known standard model sources could on average explain the presence of just 4.4 events in those two combined subsets. The excess of 13 observed over 4.4 expected counts was a sizable discrepancy, but by itself was not a real evidence of something new in the data. However, what was really startling was the weird kinematics that those events displayed. The jet containing the soft lepton had a very high energy, much higher than what backgrounds predicted. Also, the "soft" lepton inside the jet, the one spotted by the SLT algorithm, was not soft at all: many of the 13 SLT electrons and muons carried momenta in excess of 20 GeV, hence five times larger than their expected values. The other kinematic distributions also appeared quite different from expectations. In simulated top-quark samples one could not find those odd, energetic jets, containing a secondary vertex tag as well as a very energetic lepton. Those were not ordinary jets: they were *superjets*!

Kinematic Tests

The numerical excess of superjets was an *a posteriori* observation, as there was no reason to single out 2- and 3-jet events, calling them a "signal region." As such, it carried relatively little surprise value: the more places one looks in, the higher is the chance of observing some numerical fluctuation. Instead, the weird features displayed by superjet events could be argued to be genuinely *a priori*, as they resulted from the investigation of a sample isolated by independent means. This made it quite interesting to estimate the significance of the disagreement between the kinematics of the 13 events and those of standard model processes.

If you remember the discussion of the significance of the high-E_T jet excess in Chapter 9, you will recall that the arbitrariness in that procedure stemmed from two sources: the choice of what data sample to test (i.e., the E_T range where to compare data and theory prediction) and the choice of a test statistic. The same issues arose in the case of the evaluation of the compatibility of superjet events with standard model predictions. For superjets what one needed was a list of event characteristics which could describe in a complete way their kinematics, as well as a suitable statistic to test the standard model hypothesis.

Giromini put together a list of variables which could be constructed with the observed energies and directions of the bodies detected in the superjet events. His choice was driven by the desire to picture as thoroughly as possible the oddities of the sample without leaving anything out, even characteristics which could in no way be discriminating features of new physics. Similar to the choice of a wider range in the jet E_T spectrum of Chapter 9, this looked like a conservative procedure: by including variables hardly sensitive to extraneous contributions to the data in a global measure of agreement, the statistical significance of a discrepancy in the more physics-sensitive variables would be made weaker. Nonetheless, the specific choice of variables could be deemed arbitrary, thus questionable without a chance of a resolution.

The definition of a test statistic was also going to have considerable repercussions on the ensuing controversy within the collaboration. The method used by the Frascati group was a custom-made variant of the *Kolmogorov–Smirnov* (KS) test. The KS test extracts a p-value from the maximum difference between the fractions of data falling below any given value in the two histograms. Because it does not compare the contents of the bins directly, but it instead considers cumulative distributions, the KS test is sensitive to bulk differences. For the same reason, it is quite insensitive to the occasional spike (a high bin, surrounded maybe by a deficit in neighboring ones) or to *outliers*, entries in bins located far away from the regions of maximum frequency, where the distributions have otherwise died out. This makes the KS a robust test of the overall agreement of two distributions, but not the most sensitive tool to put in evidence the presence of localized differences. In that sense, the KS test could again be considered a conservative choice in the test of the superjets kinematics, as typically new physics effects are found as localized deviations in a kinematic distribution, rather than in the guise of a global shift of its mean.

Once run on the distributions chosen by Frascati, the KS test returned a set of p-values. Assuming that the kinematic variables were not correlated among each other and that the Monte Carlo provided a correct modeling of the data, the p-values were expected to be distributed uniformly between zero and 1. What instead those numbers showed was

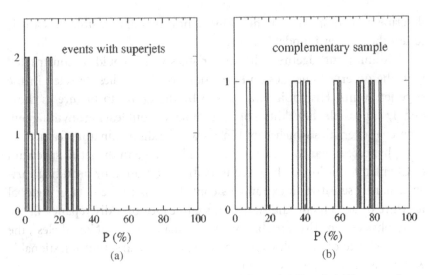

Figure 1. The histograms show the distribution of the probability of 18 KS tests, each performed on a different kinematic variable for the sample of 13 superjet events (a) and for a complementary control sample of 42 W+jets events (b) where the SECVTX tagged jet contains an SLT candidate track which is however not identified as a lepton. One expects a uniform distribution for these histograms if the data comes from standard model processes.

striking: they crowded around zero! Most of the tests signaled that the set of 13 events had a small probability of being seen. What would you think if, by throwing a die 18 times, you only got ones, twos, and threes? You would probably conclude that the die is loaded: your observation in fact may happen by chance only once every two-to-the-eighteenth times on average, i.e., less than four times in a million. Similarly, the small probabilities of the 18 tests performed by Giromini were a strong indication that the 13 events were different from the sum of SM backgrounds they were tested against (see Figure 1). Giromini found that the overall probability of the data was less than one in a million: a significant *observation* of a new physics signal!

Godparent

The analysis of the 13 events was first documented in two CDF notes in September 1997, but the collaboration did not pay much attention to them. Besides, Giromini and Ptohos had grown tired of the Top group flak, so they did not follow the meetings nor presented their work there any longer. After several months of further investigations, Giromini paid a visit to the spokespersons of the experiment: Alfred Goshaw, a professor from Duke University, and Franco Bedeschi.

One year earlier, the spokespersons had appointed a Godparent committee to review the new Frascati measurement of the top production cross section: that work had been around for a while now. The committee, led by Paul Tipton, had been reviewing quite slowly the difficult analysis, but it seemed now close to converging to a recommendation to publish. But Goshaw and Bedeschi were now going to have to deal with a totally new request: Giromini demanded to proceed to a quick publication of the analysis he had performed on the 13 superjet events. This included both the study of the kinematics of the anomalous events and a possible explanation of their weird features in terms of a new physics model.

The existence of a CDF note carefully documenting the Frascati analysis was a good start, but it was only one of several prerequisites to fulfil before one could proceed to publication of a scientific result. CDF procedures foresaw that only after a successful blessing could the authors ask for godparents and proceed to writing a paper draft. Yet, none of those additional requirements had been fulfilled by the Frascati group: they did not want to submit to the rules of the game anymore. Giromini believed that the result required a fast track to publication, because of the importance of an observation which potentially indicated new physics beyond the standard model. In addition, he wanted to avoid having to go through a Top group blessing, where chances to get stuck in the mud would be very high. In principle, the spokespersons had the power to waive the rules written in the CDF bylaws, but they were not convinced that such an extreme measure was needed.

After the meeting with Giromini, the spokespersons decided to appoint a beefed-up godparent committee which would look at the analysis of superjet events much more carefully than what typical godparent committees used to do with results under review. They chose a few experienced and authoritative figures, as well as a few young and hard-working researchers who could dig deep in the analysis details. A clear response from such a committee would be a very useful handle to decide whether the superjets paper was to be pushed through.

On the evening of October 17, 1998 I was invited to play Bridge at Bedeschi's house in Wheaton. Playing Bridge against Franco and his wife was not easy: they were a very well-trained couple, who understood each other much better than did I and my pal, Mosé Mariotti. During an early pause in the game, which saw me and Mosé already struggling, Franco enquired:

"So, Tommaso, how is your thesis writing going?"

"Well, it is mostly done by now...I think I am in good shape for defending it next January." I replied.

"But I heard that you have already been hired by Harvard to attend to the upgrade of the muon system?!"

"Right, I am in a superposition of states right now: half graduate student, half post-doc!"

"I see. So, what would you say if I asked you to spend some of your non-existent spare time on a very interesting, but complicated review task? I mean one involving real work, not just the warming of a chair." Franco asked.

"Well, sure!" I enthusiastically answered. "What is it exactly that you have in mind, though? Do you think I would be able to carry out that job?"

"I think so." was Franco's reply. "Actually, I have already discussed the matter with Goshaw. We agree that you would be a sound addition to a certain committee we are trying to put together. It is for a controversial analysis by the Frascati group. Have you by any chance read their last few CDF notes?"

"Hmm ... No, but it looks intriguing. Please count me in!" I happily replied.

"Ok, but bear in mind that it is a controversial matter, and we need people who really look into the details of their analysis. It is rather sensitive material. But you have experience with b-jets, with SECVTX tagging, with top quark kinematics ... I am sure you will do a great job."

The rest of the Bridge game went further downhill for my side; I was not paying much attention to the game anymore, as I was already thinking at the CDF note I would read that night. Franco had picked the timing of his offer well!

The next day I received an official e-mail where the spokespersons asked me to take part of the "Superjet godparent committee." I was in the company of a group of qualified and esteemed colleagues. Brig Williams was the chair; then there were Rob Roser; Brian Winer, one of the authors of the top discovery analysis; Dave Gerdes, the author of the JETPROB algorithm; and Regina Demina, a physicist who had recently joined CDF after working for its competitor DZERO. The group thus counted six members, twice as many as ordinary Godparent committees. This was a signal that the procedure would be different from the usual one leading analyses to publication. In fact, the e-mail clarified that our mandate was to assess the soundness of the Frascati analysis in depth and quickly. It also emphasized the importance of our review given the controversial nature of the claims, and explained without euphemisms that we were expected to produce real work to confirm or disprove those claims.

The first meeting of the newly formed committee was held shortly thereafter. I was pleased to observe that my colleagues took the matter very seriously. They did not express a preconception that the 13 events were actually just another kind of background previously not accounted for, or the effect of some other analysis mistake. Roser, Winer, and Williams, in particular, were old members of the Top group, and as such they well knew how to sail in the troubled waters of ill-supported claims, but they seemed open-minded about the work of the Frascati researchers.

We compiled a list of the details of the superjet analysis in need of a quantitative check. Then Brig proceeded to distribute the tasks. I volunteered to investigate the soundness of the p-value calculations contained in the Frascati CDF note. My colleagues were happy to leave that job to me, as it smelled like a lot of work. And indeed, as I set out to verify the results of the KS tests, I found out it would not be a quick job. My plan was to start by recomputing one or two numbers contained in the note by using exactly the same input data and the same method as the one used by Frascati. In case I found an identical result to theirs, I would then be able to change some of the hypotheses and explore alternative ways to answer the question of how anomalous the 13 events were. This would allow me to estimate the possible "experimenter's bias" that Giromini could have caused to his own results.

Unfortunately, getting my hands on the source code that Paolo had used for his KS tests proved anathema. Even obtaining the data was a source of pain, as he was not very collaborative. To Giromini, the desire of our committee to check his numbers appeared little short of an insult; in addition, he found it entertaining to tease me. It took me quite some time to convince him that there was nothing funny in keeping me idly waiting for his input.

As I finally obtained the data from Paolo, I started to work with enthusiasm. In a few weeks I could verify the Frascati results and perform several additional tests. I produced a CDF note detailing my results and proceeded to inform my colleagues in the committee. In the meantime, they had checked many other features of the 13 events. No smoking gun of a detector effect causing the observed phenomenon had been found. Jets looked like jets, leptons were regular leptons, missing transverse energy was well calculated, and SECVTX and SLT b-tags did not show any pathological features: the events looked genuine. In addition, some details appeared to favor the interpretation that the events were due to new physics, others instead pointed to some weird origin related to the geometry of the CDF detector. In particular, the 13 events showed an unexplained left–right asymmetry. In many of them, the W decay lepton was emitted in the direction of motion of the proton beam, toward the edge of the central region of the detector. The speculation that what was causing the events was a new physics process which distinguished protons from antiprotons, thus creating a left–right asymmetry in the observed final-state particles, was one really hard to swallow. A foundational theorem in particle physics, called "CPT," would have to be abandoned for that to happen.

There laid a real quandary: even if one admitted that the events were due to some exotic new physics process, one would be forced to accept that the new physics was really like nothing ever dreamt by even the most daring theoretical physicist. However, Giromini had put forth a tentative model involving a process which could explain, at least in part, the phenomenology of superjet events. His explanation would become a central issue in the superjet affair.

Chapter 12

Scalar Quarks?

The 13 superjet events had only started to stir up trouble in CDF when Giromini came up with something even harder to deal with. There had, in truth, been warnings that their anomalous observation would be the beginning of a bigger story. A significant portion of Frascati's CDF note 4348, the main documentation of the superjet analysis, dealt with the description of a new physics mechanism which could be the source of the observed signature. The note also included some preliminary tests of how that hypothesis could fit the numerical excess of W+2 or 3 jet events as well as explain their kinematics. The Superjet godparent committee had been specifically asked to stay clear of that part of the documentation, at least in a first phase of the review, and only focus on ensuring the correctness of the analysis.

A Supersymmetric Bottom Quark

The hypothesis contained in the interpretation section of CDF note 4348 was startling. The superjet events could be explained by the production of a new supersymmetric quark. This *squark* would be endowed with some characteristics similar to those of bottom quarks, but a few others very different. A bottom squark of mass in the 3–4 GeV range could fit the bill if it had a lifetime a bit shorter than a picosecond and a decay always yielding a charged lepton. The b-squark, produced together with a W

249

boson, would give rise to a jet easy to b-tag with the SECVTX algorithm. Also, the jet would *always* contain an electron or a muon, which the SLT algorithm could see. That hypothetical feature, distinguishing the squark jet from regular bottom-quark-originated ones, could explain the excess of SLT tags in SECVTX-tagged jets. Furthermore, the intrinsic properties of supersymmetric quarks could account for the unusually large momentum of the charged lepton in the superjets and the other observed characteristics.

Indeed, the pages of the Frascati note which I had read with more interest from the start were exactly those containing the above interpretation. After looking at the odd kinematic distributions of the superjet events, and noticing how small the probability of observing a similar effect was, it would have been strange to feel no attraction for the interpretation part of CDF note 4348. But skepticism had to come first: it was an obligatory reaction for any physicist endowed with a minimum of common sense and a bit of knowledge of the history of particle physics — studded as that is with entertaining examples of anomalies that produced hasty claims, later awkwardly retracted.

Hundreds of questions crowded my mind: why do b-squarks show up in the company of W bosons, but are not seen in other kinds of events? What is the production mechanism? Why is the superjet always so energetic? And above all, how on Earth could such a lightweight new particle have escaped detection in the score of lower-energy experiments that preceded CDF in the investigation of particle collisions? I did not believe that there could be a light b-squark in our data; it had to be a spurious signal, and I thought I could prove it to Giromini. Yet, as I visited him in his office one afternoon, clasping a rolled-up copy of his 60-page note as a billy club, my cockiness did not last for long: Paolo had a plausible answer to everything.

The issue which had fueled my skepticism the most was the absence of any hint of a light supersymmetric quark in electron–positron collider data. If squarks exist, they can certainly be pair-produced in electron–positron collisions. This is granted by their non-zero electric charge: owing to that they must couple to the electromagnetic field. A sufficiently energetic photon may thus materialize a particle–antiparticle pair of those bodies. What I had not considered, however, was that unlike standard model fermions, that carry an intrinsic spin of 1/2, squarks are scalar particles of zero spin. That property strongly reduces their pair production by a virtual photon, making the process hard to detect in lepton collisions. I learned that piece of wisdom the hard way from Paolo, who enjoyed pontificating on my scarce understanding of the phenomenology of supersymmetry. Apparently, it was not so easy to dismiss the hypothesis that superjets were due to bottom squarks: that tentative explanation held water tightly enough. I thus meekly unrolled my billy club and went back to occupations more fitting for my limited skills.

The Second Anomaly

Giromini soon took the consequences of his hypothesis of a SUSY quark to the logical next step. If the squark decays to a charged lepton every time, as superjet events suggested, a signal of its presence must show up in the inclusive lepton samples. Those were the datasets which had been the basis of my search for a $Z \rightarrow b\bar{b}$ decay (see Chapter 10).

Giromini was quite familiar with inclusive lepton datasets: his group had used them to determine the SECVTX scale factor, the number on which their precise measurement of the top production cross section was based. For that task, a very large sample of Monte Carlo events containing two jets, one of which included a lepton, had been produced by the prolonged and painstaking effort of Fotios Ptohos. Fotios had drained for months the CPU of the large cluster of Frascati processors with his simulation jobs.

The comparison of the number of SECVTX b-tags in data and simulation, together with the knowledge of the efficiency to b-tag bottom and charm jets, allowed to compute the fraction of jets in the data originated by bottom quarks, charm quarks, and lighter quarks or gluons. That information could then be used to predict the number of additional leptons in the data besides the one triggering the event collection. The contribution of an unpredicted extra source of leptons, such as the pair production of squarks with 100% decay probability to electrons or muons, would thus stick out like a sore thumb.

Of course, there are subtleties in such a statistical analysis. For example, jets originated by light quarks or gluons may occasionally also be tagged by SECVTX or contain identified leptons. The number of those spurious signals must be sized up precisely. One also needs to assess the possible correlation between the probabilities to find SECVTX b-tags and to find leptons in jets. The study of the rate of spurious leptons and vertex b-tags had already been performed by Frascati for the top cross-section analysis. The same goes for their hypothetical correlation, which was a critical point in the superjet analysis.

The result of fitting the observed number of SECVTX b-tags in the data was a very precise estimate of the sample composition: about 40% of the jets were understood to be due to bottom quark decays, and 10% to charm quark decays. The fit was able to correctly predict the sharing of vertex b-tags in the lepton-side and away-side jets; hence, one could be confident that the fractions were estimated correctly. However, in the subset of events containing a second lepton, either in the lepton-side jet or in the away-side jet, things went seriously awry: the number of those multilepton events significantly exceeded expectations. The data showed a 20% excess of events with two leptons, which grew to 45% if the additional lepton was sought in the away-side jet. A second anomaly!

While the numerical excess was more prominent in events with one lepton in the lepton-side jet and one in the away-side jet, it was the study of events where the same jet contained two leptons that provided the strongest case for the presence of supersymmetric quark decays in the data. Those events offered a chance to study the relative kinematics of the nearby leptons and produce comparisons with expectations from all known standard model sources. In that subsample, the excess events were found to preferentially contain lepton pairs with a small angular separation. In the scalar quark hypothesis, such a configuration could be understood as due to the lighter mass of the b-squark with respect to that of the bottom and to its different spin. This could be considered an additional confirmation of the new physics explanation!

The number and distribution of excess leptons in inclusive lepton samples could be explained well if one assumed the presence of a 15% contamination from bottom squarks in the data. The best fit was obtained by adjusting the unknown lifetime of the bottom squark to 0.6 picoseconds, and the mass to 3.6 GeV. It looked too good to be true — but was it?

The Godparent Report

The CDF note detailing the analysis of the inclusive lepton samples was produced by Giromini's group in March 1999, while the Superjet godparent committee was struggling to reach a consensus on how to proceed with the superjet events. It was definitely a bad timing: the new anomaly made things significantly more complex. Until then, a lean publication of the 13 events which could describe their strange observed characteristics while omitting any claims, and delegate to future articles any possible interpretations of the observed effect, had been an option. Giromini had kept rejecting such a plan, but it had remained the goal of our godparent committee. Now, however, the new analysis had crept in, blowing that plan off the table. Giromini was not helping in the search for a solution, since he claimed that the second anomaly demanded a rapid publication. He warned the spokespersons that in case of further delays he would go ahead and publish an article describing his results irrespective of the consensus of the collaboration. Of course that action would cause his immediate expulsion from CDF, but the spokespersons were not in a position to ignore the threat. If he published the analysis by himself and later the bottom squark signals turned out to be genuine, Giromini would be recognized as the author of a monumental discovery overlooked, or worse, hindered by his colleagues. CDF could not take such a risk. The spokespersons could only ask the godparents to speed up their review and indicate whether the results should be published.

Our verdict was a split one. Dave Gerdes and I believed that the 13 events and their odd features had to be published in some form. They represented a very significant, unexplained effect, one which an extensive review had been unable to explain away. It was a possible indication of physics outside the standard model! Hiding such a potentially relevant observation from the rest of the community of particle theorists and experimentalists seemed a scientifically unsound decision. It was potentially even a hindrance to the progress of science. CDF owned its data, but eventually, the data were also the property of mankind, no less than all the other bits of knowledge accumulated through centuries of thought, investigation, and government-funded experiments.

The other four members of the committee did not object to the above concept, but they gave more weight to the reputation of CDF. They felt that CDF had to keep a high-quality standard for its publications. A paper ill-motivated in its premises and inconclusive in its findings was in their opinion below threshold. They did not object to the decision to publish the superjet events, but neither did they recommend it.

On May 27, 1999 the Superjet godparent committee detailed the results of their review in CDF note 5015 and presented it to the spokespersons. After a detailed description of the studies performed to verify the work of the Frascati group, the document explained in its concluding section that nothing wrong had been found with the events or the analysis. Indeed, the probability of observing the weird kinematics of the 13 events, if the data came from standard model processes, could be assessed in the one-in-a-million ballpark. As the events could be a signal of new physics, the committee suggested that CDF should seriously consider publishing them. However, the committee members stressed that they were not unanimous on whether to indeed publish the results straight away or to perform further studies. The message which transpired was a fright to publish results that were not "fully understood" — granted that a full understanding of a physics phenomenon is anything but a low-hanging fruit. In any case, Goshaw and Bedeschi had now something to cling on in order to solve the controversy, which was threatening the unity of the collaboration as well as significantly diverting resources and manpower from the goal of getting ready for the impending start of Run 2.

A Third Anomaly: the Dimuon Bump

In the meantime, the Frascati group kept adding gasoline to an already vigorous fire. Ignoring the continuous opposition of their colleagues to their analysis, or maybe trying to win a larger consensus, they sought to provide additional evidence of their new physics interpretation. To Giromini, the excess of events with multiple leptons was speaking loud and clear about the existence of a scalar bottom quark. He had measured the properties of that tentative new object using the kinematics of lepton pairs in the same jet, but whole books remained to be written on the phenomenology of the unexpected particle.

The next logical step was not a hard one to make: supersymmetric or not, quarks withstand QCD interactions and usually bind into colorless hadrons. Given the 3.6-GeV mass hypothesized for the bottom squark, the exotic hadrons containing it are unstable and must decay into lighter bodies. If the Frascati hypothesis were correct, a new meson made by a bottom squark bound to an ordinary quark would predominantly decay to a lepton, a supersymmetric neutrino, and a charm quark. Unfortunately such a decay would not be fully reconstructable, because the supersymmetric neutrino would escape unseen, carrying away energy and momentum. A hadron containing the squark which could give hope to be observed as a bump in a mass distribution (the most convincing signal of a new particle one can pull off) was instead a *vector meson*, a particle like the J/ψ or the Υ. Those bodies are called "vectors" like the weak bosons as they have an intrinsic spin equal to 1.

A squark with mass in the 3.6-GeV ballpark bound to its own antisquark should form a 7–8-GeV particle. As squarks have *zero spin*, two such bodies may create a vector meson only if they orbit around one another with one unit of relative angular momentum. The total angular momentum of a composite system is in fact the vectorial sum of the spin of the constituents and their relative angular momentum. Two spin-1/2 particles may naturally make a spin-one bound state if they have no relative angular momentum, as in the case of the J/ψ meson; two spin-0 particles instead need to provide the extra unit through their relative motion. And since a non-zero orbital angular momentum between the constituents generally makes bound states harder to produce, this implied a smaller chance of the two squarks to bind, i.e., a smaller cross section. But maybe the number of bound states produced would still be large enough to be seen.

As with any other unstable particles, the chance to observe vector mesons depends on the way they decay. J/ψ and Υ mesons are easy to spot in hadron collisions when they decay into a pair of muons. This *dimuon* signature is advantageous because it is easy to reconstruct and there is comparatively little background, so even a small signal can be seen with clarity.

We may tentatively call E (for Epsilon) the hypothetical particle made by bottom squark pairs, following Giromini, who picked the name to hint at the predicted smallness of the resulting signal. Epsilon is, in fact, the Greek letter which mathematicians usually employ to describe arbitrarily small numbers. To search for Epsilon, it was natural to use the sample of events containing signals of two muons, a dataset which the CDF detector collected with a dedicated trigger. Giromini knew that the E particle should have a mass lying somewhere between 6 GeV and 8 GeV, i.e., about twice the mass he had estimated for the supersymmetric quark. He also knew how to compute the number of such bodies expected in the data. Even avoiding any major arbitrary assumption about production modes and phenomenology, it was possible to relate the production rate of Y mesons and E mesons, accounting for the different mass and physical properties of the two particles with the help of theoretical calculations. This way, by suitably rescaling the number of observed dimuon decays of the Y meson, one could predict that the dimuon dataset should contain about 250 E decays. That was a small signal, but maybe a detectable one!

In order to observe a 250-event bump amidst the huge background of non-resonant muon pairs a careful selection was needed, which could clean up the signal region of the mass distribution as much as possible. The Frascati team imposed tight requirements on the quality of muon candidate tracks and on their kinematic characteristics. They could show that the same requirements did effectively remove backgrounds under the towering 10,000-event Y signal, which laid at just 2.2 GeV higher mass in the same distribution. This proved that they were thus not made up or tuned *a posteriori*. And amazingly, after that selection, it appeared possible to indeed fit a resonant-like structure in the dimuon mass spectrum, at a mass of 7.2 GeV, and with a size matching the expectation of the scalar quark model (see Figure 1)! As I report this, 15 years later, that observation still mystifies me. Fluctuations of that kind do occur in the data, but a better explanation of the phenomenon is that some playful God decided to confound the CDF experimentalists, producing a string of anomalies that could be interpreted as a coherent picture of new physics.

The scrutiny of the dimuon data by an independent party now became urgent business: the new signal appeared to confirm the scalar quark hypothesis, and Giromini felt even more justified in his hurry. According to him, what was needed was to bypass the Top group, by forming a panel of reviewers that were not prejudiced against his analyses. Such a panel could hopefully reach a consensual view on the publication readiness of his results.

In response to the emergency, Goshaw and Bedeschi put together a new structure of committees to handle the expanded set of controversial results. A group of four reviewers was appointed in a "Scalar Quark committee" chaired by Brig Williams, to verify the soundness of the inclusive lepton

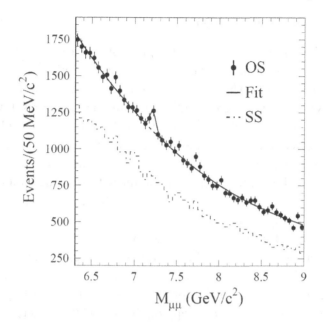

Figure 1. The mass spectrum of selected muon pairs (points with uncertainty bars) in the 6.3–9.0-GeV region is fit to a falling exponential with the addition of a 7.2-GeV Gaussian signal of width compatible with detector resolution. The dashed histogram shows the distribution of muon pairs of same charge (labeled "SS"), which show a smoother distribution. Reprinted with permission from *Phys. Rev. D* 72 (2005), 092003.

analysis and the real significance of the dimuon bump. And a second group, called "Oversight godparent committee" and chaired by Bill Carithers, was charged with coordinating the review of the three anomalies, collecting a report from the Scalar Quark committee, and

> "investigate explanations of the superjet and scalar quark phenomena in terms of unusual detector effects, untested standard model backgrounds, or proposed interpretations in terms of scalar quarks or other phenomenological models.[...]"

If the measurements were found to be correct, the committee was then to

> "make recommendations to the collaboration about the manner in which the observations should be published."

Bottom Squarks in LEP Data

Between the summer of 1999 and the spring of 2000 the battle kept raging between the Frascati group and the CDF management. There were now multiple open fronts. The one where things appeared more favorable to Frascati was the new top cross-section measurement. That result was now well documented in a bulky and detailed article, whose path to publication was now cleared out thanks to the positive attitude that Paul Tipton's committee had started to show. The reviewers were now comforted by the precise explanation produced by the authors for the high value of the SECVTX scale factor. Tipton was a veteran of top quark studies; his word had a serious weight within the Top group.

On a second front, the superjet analysis was suspended in a sort of limbo: after the split verdict of the godparents, progress had been almost null. Most of the action was centered on trying to converge on a paper draft which could be acceptable for CDF, with Giromini resisting the various proposals of reducing or dropping the new physics interpretation in the text. That article was also kept on hold by the fact that it needed to cite the improved determination of the SECVTX scale factor in order to produce a precise background estimate in the superjets sample. Because of that, it could arguably be published only after the top cross-section article, or simultaneously with it. And finally, the third front was constituted by the squark signal in the inclusive lepton samples and the dimuon bump at 7.2 GeV; these were the subject of internal scrutiny by the two new godparent committees, but progress was again very slow.

Then, during the spring of 2000, news of the possible presence of a squark signal in CDF data hit CERN. Although I investigated the matter in some detail, I was unable to ascertain who leaked the information to the competitors. In a strict scientific sense, the idea of informing the CERN experiments was sound: if a light supersymmetric quark existed, it was certainly possible to produce it in particle–antiparticle pairs via electron–positron collisions. LEP II was now producing such collisions at energies up to 208 GeV. The signature to look for there was clearly that of pairs of back-to-back jets containing multiple leptons. Any sort of confirmation of the scalar quark signal from LEP II, even an unofficial one with less than discovery-level strength, would mean a lot for the publication path of the CDF analysis. On the other hand, the leak could potentially strip CDF of the paternity of a monumental discovery!

The first work on the bottom squark signal was done by Mario Antonelli, an Italian physicist working for ALEPH, one of the four LEP collaborations. He prepared a simulation of electron–positron production of bottom squarks with the characteristics tentatively inferred by the CDF analyses and spun the available ALEPH datasets in search of dijet events with leptons. Amazingly, a quick-and-dirty analysis evidenced a surplus of events in the selected sample with respect to background predictions: a 3-sigma excess of dijet events where both jets contained a soft lepton! Antonelli was intrigued, but many details required a more careful treatment. He thus met with Paolo Giromini, who had traveled to Italy at the end of June; they exchanged information on the signature, such that Antonelli could refine his analysis. Then Giromini informed the CDF review committees that LEP was also seeing his signal.

There followed a few frantic weeks. On June 29, a meeting between members of the Oversight godparent committee and the Scalar Quark committee members took place to assess the situation. The chance that CDF would get scooped on the discovery of the century was now real: the three sigma of the ALEPH signal could easily grow to discovery-level significance if the four LEP experiments decided to combine similar results. In CDF, the superjet paper draft was now in a reasonable form, having been purged of some of the steepest claims it originally contained. It now had bare hints at possible new physics interpretations in place of a whole section on the phenomenology of bottom squarks. On the other hand, the reviews of the inclusive lepton excess and the dimuon bump were still in the jungle, although nobody had found a mistake in those analyses in the course of the past year. The Scalar Quark committee members assured that they were close to issuing a note which described their work. Given the situation, the two committees agreed that it was important to inform the collaboration and make an attempt at pushing two articles to publication. A letter was prepared which would accompany the release for internal review of the two drafts. The letter included the following statements:

> "The Frascati group performed a detailed analysis of W + jet data containing heavy flavors tagged by SVX+SLT (the so-called "superjets").

The analysis has been extensively checked in 1997–98 and then later in 1999 by two different godparent committees. The analysis finds a significant excess of W+2, 3 jet events [...] that cannot be described in terms of Standard Model physics.

The Frascati group has also performed a thorough comparison of heavy flavor yields in low-Pt inclusive lepton data, finding significant excesses [...] as well as inconsistencies in some kinematical distributions which could be explained by the production of a scalar quark [...]. Furthermore, an analysis of dimuon data finds a bump in the invariant mass distribution [...] which would be consistent with the interpretation of the anomalies in the inclusive lepton and W+jet samples.

Two PRD drafts detailing these analyses have been produced and have been under extensive review by members of our godparent committee during the last three months.

Although close, we are not completely finished with our review. However, there are added reasons why we feel we need to urge the collaboration to carefully read these PRD drafts. Unfortunately the results of these analyses have become widely known outside CDF, resulting in LEP analysis groups pursuing this same possible signal, with unofficial word of a significant excess found already. [...]"

In the meantime, CERN was moving forward. At the beginning of July, ALEPH informed the other experiments of the ongoing work of Antonelli, providing details on the signature to look for and a pointer to the generated squark simulation. Each experiment assembled a small team to work on the bottom squark signature. However, ALEPH hung on to its lead: at the July 20 "LEPC meeting," the forum where the four experiments periodically informed each other of their latest results, it was the only collaboration presenting the status of its bottom squark search.

The ALEPH presentation was given by Gary Taylor. It encompassed all the searches which were ongoing in the freshly collected LEP II data and of course focused on the Higgs boson search, which was now starting to show some excess at 115 GeV. That was the beginning of the saga of the tentative Higgs boson signal which dominated the scene at CERN for two more years before the effect got archived as a fluctuation. After carefully discussing the Higgs candidates and other SUSY searches, Taylor showed four slides summarizing Antonelli's results on the squark search. The first one in the set was titled "Light sbottom search — as a result of a communication from CDF." After a definition of the object being sought, and one further slide describing the experimental selection aimed at enhancing a possible CDF-like signal, the speaker showed a table with the expected and observed events at each center-of-mass energy probed by the LEP II data, from 161 to 205 GeV. By combining the data, the number of jet-like events containing opposite-sign leptons was 56, when 33.6 were expected from standard model sources. On the complementary sample of events with same-charge leptons, which the signal simulation predicted to be almost free of contamination from bottom squarks, the agreement between data and standard model prediction was instead very good. The last slide summarized the preliminary ALEPH result: the probability of the excess of 56 over 33.6 expected events was of 2.5 parts in 10,000. Once one included a conservative 10% systematic uncertainty on the background rate, the *p*-value increased to two-thousandths — still a nearly 3-sigma effect.

As Taylor gave his report an important international conference, the "Recontres du Vietnam," was already ongoing in Hanoi. Monica Pepe-Altarelli was due to report the new LEP II results there. She had delayed her departure from CERN precisely for the purpose of collecting the latest information at the LEPC meeting. Three days later, she could finally show to her audience, among many other results, the intriguing ALEPH signal of bottom squark pair production.

The divulging of the 3-sigma ALEPH excess in a public arena created some turbulence within that collaboration, as Antonelli's analysis was preliminary. It had been only approved to be discussed within the LEPC group and was not meant for external distribution. Meanwhile, one day after the LEPC meeting, a couple of shortcomings were found on the results of that search. First of all, the simulation of bottom squark production was understood to have been produced with a wrong setting in the Monte Carlo program. This setting filtered out all events that featured hard final-state radiation. That invalidated the prediction of some of the kinematic characteristics which the signal would show, making the ALEPH data selection no longer optimized to evidence it. And second, electrons in the jets had been selected with looser identification requirements than standard ones; the corresponding estimates of backgrounds due to spurious electrons were statistics-limited and perfectible.

Once those shortcomings were fixed, a sizable reduction in the scalar quark excess was observed.

The ALEPH physicists were of course not happy of having presented a spurious signal as a potential ground-breaking discovery at an international conference! Furthermore, Monica Pepe-Altarelli's talk offered to the competing LEP experiments the chance of quickly showing that they were not seeing anything in their own data, disproving the anomaly found by Antonelli. This happened at ICHEP 2000, the biennial International Conference on High-Energy Physics, which was held in Osaka at the end of July. The OPAL collaboration reported searching for the light bottom squark. They found 15 events when 20 were expected from backgrounds: no excess at all. Mario Antonelli was also present in Osaka. He gave a talk where he discussed various results of supersymmetric searches at LEP; however, he made absolutely no mention of the bottom squark search, as now explicitly requested by the management of his experiment. A few days later, the CDF expert of SUSY searches John Conway did nonetheless show one slide on the matter in his plenary talk. The slide mentioned the tentative signal found by CDF, the spurious ALEPH signal, and the non-confirming result by OPAL. In turn, John's adamant reporting of the situation also raised some eyebrows in CDF and at CERN. The bottom squark analysis was not blessed by CDF, but John's disobedience to the CDF bylaws was soon forgotten in the convulsive situation.

To complete the picture, it should be mentioned that one month later, at the LEPC meeting of September 5, DELPHI also claimed to see no bottom squark. No further word of that signature was made by the other LEP experiments, and everyone's attention settled for good on the few wannabe Higgs candidates at 115 GeV. Finally, the phantomatic squark signature was also searched by the CLEO collaboration in low-energy electron–positron collisions. They reported a negative result in a preprint released on October 14.

Irrespective of the true or spurious nature of the reported signals, the matter in CDF remained red-hot throughout the summer, as the two draft papers describing the Frascati superjet and scalar quark analyses were finally distributed for internal review. The collaboration was thus officially being asked to provide feedback on those documents. By mid August, the Frascati group and the godparents were submerged by scores of long email messages containing strings of comments and detailed criticism to every sentence. The problem was that the most negative responses, which came from groups historically most adverse to Giromini's works, called into question issues that according to the godparent committees had been addressed satisfactorily by the authors. Giromini could thus retreat into a simple "no comment," refusing to address the matter in detail. The institutions in turn felt authorized to complain with the spokespersons that their comments were not being addressed. The spokespersons finally scheduled a special meeting with the aim of informing the collaboration of the situation and verifying whether some form of consensus could be reached on how to proceed.

The Lightning Rod

Avi Yagil's office in the CDF trailers was only two doors away from mine. Avi used to arrive at 11 in the morning and leave at 12 for lunch; then sometimes he would be back for a couple of hours in the afternoon. I knew his office schedule well because I had learnt to distinguish his pace every time he walked the long corridor behind my back. He had a peculiar way of walking, a habit he had probably acquired during his military service in Israel. It produced more noise on the carpeted floor of the portakamps than you would think possible. His habit of speaking with a loud voice and his friendly interaction with the colleagues he met in the trailers also helped tagging him as he approached. I had come to know very well what Avi thought about the Frascati analyses: his opposition to the revised top cross-section measurement, for instance, had been fierce. I liked to prod him when the chance arose, however, as his opinion was always interesting to hear. Avi, on the other hand, knew that as member of the Oversight godparent committee I was supporting Giromini's request of a quick publication of the results; yet, on the whole he seemed to have a good opinion of me as a physicist.

I stopped him in front of my office on a beautiful June morning, the day before the special meeting, as he arrived fresh as a rose.

"Hi Avi, have you got a minute?" I asked him.

"Sure, baby. What do you need?"

He was well-dressed as usual, a sports T-shirt carefully tucked under the freshly-ironed trousers, mocassini on his feet, no socks.

"Well, I would love to hear your opinion on the issues that will be discussed tomorrow. Come in, take a chair."

Avi stepped in my office, but he did not sit down.

"You know my opinion. I'm telling you, that's serious junk. Those guys are monkeys! Monkeys dealing with heavy machinery. Very dangerous."
He explained, waving his hands.

"Come on, Avi. What's the heavy machinery?" I inquired.

"Claudio and I worked for years to understand soft-lepton tagging. We put together an algorithm that is not for everybody to tamper with. You know all this stuff. SLT works wonders, I'm telling you, but you have to know what you are dealing with. In the beginning we sat together and I gave Paolo all the instructions on how to use the SLT. But they do not really have a clue. You put artillery in the hands of monkeys, then you tell me what you get."

"You can't be serious." I replied. "If you had a complaint on the way they estimate backgrounds in the SLT-tagged W+jets sample you should have told us months ago! I think you were explicitly involved in the review process at some point."

"No, no, the estimates are good, Tipton's committee has checked them ok. But all this issue with the scale factor is totally fucked up. Who needs another scale factor!" said Avi.

"But I bet you've heard about the result of Paolo about the origins of the difference from unity of the scale factor," I tried to argue. "Inserting a new decay table in the Monte Carlo, the other studies on the data, finding secondaries from nuclear interactions in the silicon …."

Secondary particles are created when an energetic hadron hits a nucleus of the detector material. An incorrect modeling of the rate of such nuclear interactions may affect the estimate of spurious SECVTX b-tags, because secondary tracks have a broader distribution in the impact parameter distribution, which causes them to produce secondary vertices not due to the decay of heavy-flavor particles. I continued:

"They showed they can explain away a good part of the discrepancy from unity of the scale factor. Don't you believe that the cross section for top production is lower than what we have published?"

Avi was adamant: "Listen, I don't give a shit about the top cross section. The issue is another one here. They want to prove they have found some new physics in W+jets data, and they need the scale factor higher than one to make their case stronger."

That was of course true: if you observe an excess of events and you can explain away fewer of them as being due to top production, you are left with a bigger, more significant effect.

"But we are scientists," I argued. "We have to revise our measurement if we think we have a better number. Let's keep the two things separated. I think we should let Paolo publish the new top cross section. Then take stock, and see how things stand ...."

Avi interrupted me. "I don't want another top cross section out. We have to stop this circus. It's not a matter of scientific integrity."

"But do you really think the scale factor is one?" I inquired.

"No, it's probably ok. It's 1.2, whatever. That is not the question!" was Avi's reply.

"Then let's let it out, why not?" I argued.

"It's not going to happen. They have to be stopped somewhere."

Seeing that Avi was stubborn on that issue I changed topic.

"Look, I have checked the Kolmogorov tests of Giromini. They are okay. What do you think of the kinematics of those events, then?" I asked.

"Baby, those events have been around for years. Come on! We have known since 1994 that we had some funny events there. Four years ago they tried to sell us that they'd found the Higgs with 10 odd events in the 2-jet bin. Then they changed their minds, and it was charm plus missing energy in 50 SLT tags. And now they've settled on 13 multiple tags."

It was hard to deny that Avi had a point there. Over the previous few years, the Frascati group had produced three different claims. All of them were centered on subsets of data produced by applying slightly different selections to that same sample of W+jets events. While I was seeing that sequence of studies from the positive side — a continuing investigation of the origin of observed oddities in the data and a search for possible explanations, after something odd had been spotted — Avi perceived it as a proof of unscientific behavior, and a threat to the reputation of the experiment. In CDF, he was not the only one who felt that way. I conceded him the point:

> "Ok, ok. I do not deny that. Still, I believe those events are weird enough that they should be published. What do we lose? Imagine if a bright theorist comes up with some new model and … ."
>
> He interrupted me again. "No way, baby. CDF will not publish those events. I am going to see that this does not happen."
>
> "Ok Avi, then I guess we'll have fun tomorrow … ."
>
> "Yeah. You bet. See you, buddy."

Yagil had been sincere: he did not have anything concrete to object to, with the exception of the scientific method allegedly used by Giromini, but that alone made all the difference to him. His was a good representation of the feelings of the American lobby in the Top group. In addition, they were disturbed by the fact that the result they had recently published would be revised by Frascati, and they found it wholly unacceptable that Giromini would win his battle with the superjets. It was for sure going to be an interesting meeting!

On the following day, the collaboration responded to the critical situation with a large attendance. Nearly 100 collaborators gathered in the CDF theater, the large room recently realized by remodeling part of the second floor of the CDF building. In addition, a dozen universities and research groups were connected by videoconference. The agenda included a short introduction by the co-spokesperson Al Goshaw, followed by a presentation of the analysis by Giromini. Then the floor would be open to discussion on how to proceed with the review of the papers.

Goshaw gave a brief reminder of the history of the superjet analysis and a summary of the recent events which had led to the calling of that special meeting. Then Giromini got on stage. His presentation, like others I had seen before, started clear and got murkier and murkier. His analysis was too complicated to be described in sufficient detail in a 30-minute talk, even for an audience well trained on the complexities of SECVTX and SLT tagging. Moreover, his slides were very dry: they contained only graphs and very little or no accompanying text, with no hint of what conclusions were to be drawn from them. Paolo's style of presenting "bare facts" had the apparent aim of allowing listeners to make up their mind by themselves. This, however, confused more than it clarified. The few who had studied his documents beforehand were able to follow his talk and understand his points; all others were left with a feeling of mounting frustration. Probably, the reason for the cryptic nature of his talk laid in Paolo's conviction that most of his listeners stood no chance of understanding his analysis anyway! But if it was a voluntary choice, it was a shot in his own foot: his listeners, even those who would not understand the merits of his work, would be the ones who would ultimately decide what to do with it.

And then there were the interruptions, continuous and unforgiving. I tried to follow the discussion silently, but once the issue of Kolmogorov tests arose, I could not stand hearing over again objections based on issues we had resolved one year before. As Paolo showed his p-value graphs Morris Binkley, who had performed some KS tests on his own using Paolo's data, argued that he had found different results. Asked to better explain his checks, Binkley descended the few steps of the CDF theater from the chair he had occupied in the last row, and took the floor to show two slides he had brought along. I respected Morris as a scientist and I thought he was a great guy — I loved his smile and his constant cheerful attitude. Yet, as I saw what he wanted to discuss I got upset: I did not know whether to laugh or cry. Morris had been using the Kolmogorov–Smirnov function as computed by HBOOK, an outdated program used to produce and analyze graphs and histograms. The KS test in HBOOK was known to contain a bug, which frequently caused it to return bogus results. Apparently, though, Morris was unaware of that fact. I felt compelled to interrupt the interruption:

"It puzzles me to see the significance of Kolmogorov tests discussed in this arena over again, months after the godparents produced two documents where they discuss the matter in detail and confirm Paolo's findings. I would also like to point out that, judging from the graphs he has shown, Morris is using the DIST routine in HBOOK, which does not return a correct probability estimate with normalized distributions! This problem with DIST has been confirmed by the authors of the package. I would like to encourage all of you to read the CDF note we have produced on the significance tests. I understand it is a lot of material to digest. But we should not continue discussing objections of this kind, which disregard the past work of authors and reviewers, if we want to make progress."

Morris begged to differ. "I have used HBOOK without a problem for similar tests in the past. And I think that one of the objections highlighted by the distributions still stands: it is clear that we are dealing with a few outlier events in the superjet sample, which may be driving the small significances the Frascati group is quoting. Maybe removing the events in the tail of the distributions could show that the disagreement with simulations is not that bad."

"But that is exactly what the KS test is designed to be insensitive to, Morris." Giromini stepped in to object. "This matter has not been understood well enough here, it seems. If you took the outliers in each distribution and moved them around, the probability would not change! This is exactly the reason why we chose the KS test in the first place: it is sensitive to the bulk of the distribution, not to the tails."

This remark by Paolo generated a further heated discussion, fueled by the apparent confusion on the way the KS test really worked. Digesting what Paolo meant took a significant part of the time originally allotted to his talk. But then things escalated, as Tony Liss introduced a different objection.

"It does not matter whether the KS test is sensitive to the bulk or the tails," he argued. "You are showing a combined probability which totally neglects the correlation between the variables. But you are using a score of variables! There is certainly a correlation between them, and it must affect the significance."

"Sorry Tony, but the matter has been studied extensively" I again interjected. "Pseudo-experiments show that the significance does not change appreciably if you take correlations into account. Besides, that is also the outcome of tests performed on the *complementary sample*."

I was being drawn more and more in the discussion, because of the studies I had performed. Giromini did not mind if I answered the criticism on Kolmogorov tests in his stead. I felt the burden to defend his results, which I had spent months to checks. My mandate as a member of the Oversight committee was to defend the analysis and see it through to publication, once I was convinced of its soundness. Some of my collaborators, however, perceived me as a defence lawyer of Giromini. I was seen as an annoying obstacle in their attempt to close the issue without a publication. As for the complementary sample, it was a sample consisting of W+2 or 3 jet data with a SECVTX-tagged jet which did *not* contain an SLT b-tag. Those 42 events carried the very same biases of the superjet dataset: they were by construction as similar to the other 13 as possible, while not containing the allegedly tell-tale SLT lepton. That made them an invaluable tool with which to determine the soundness of whatever test was performed on the smaller set of 13 superjet events. The p-values of the KS tests in the complementary sample is shown on the right in Figure 1 of Chapter 11.

"But the list of kinematic observables considered by the authors is ridiculous!" objected another member of the Top group lobby. "It includes several variables which cannot *possibly* be due to new physics. The position of the primary interaction vertex along the beam axis, the pseudo-rapidity of the primary lepton, and other variables must be insensitive to new physics."

The American questioner was concerned with the fact that several of the 13 events had a z-vertex which laid on the same side with respect to the center of the detector. This was an odd feature which could not be justified by any physical effect, standard or new. He was also bothered by the fact that a few of the 13 primary electrons and muons making up the W decay signal in the superjet events were concentrated at a particular pseudo-rapidity, corresponding to the edge of the central detector. He continued:

"The authors are showing that they have found an effect which *cannot* be new physics, because those characteristics can only be due to a detector effect! Basing a significance number on such characteristics is meaningless."

"This is not quite correct." I again felt the need to intervene. "The godparents have produced a document which explains that even if one selects five very well-motivated kinematic variables, ones that are *a priori* a good gauge of physical effects, the significance remains very much above four standard deviations. What the authors did when they chose their list of variables was to define a complete set of observable quantities, one which would not sound like an advantageous *a posteriori* choice for supporting their claims. They did this to pre-empt the criticism of the arbitrariness of their choice!"

Goshaw tried to bring the discussion back on track. "I would like to remind everybody that we are here today to decide whether we should proceed and publish the superjet events, or just continue investigating their nature. I need to ask you all to make an effort to stay focused on the issue at hand."

Avi stepped in: "Exactly, Al. I think it is clear to all of those present here who have been around in CDF during the last three years that the authors have been playing with the W+jets sample for far too long. If you search for a discrepancy for long enough, you are bound to find one. Had we left them publish anything they produced, we'd now have a Higgs, an exotic new physics process producing a charm-quark excess, and who knows what else."

A colleague added: "I have to concur. The superjet search is certainly not an *a priori* one. I think this is an excellent search to perform with the data we are going to collect in Run 2. With new data, and with event selection strategy and sample definition perfectsly frozen, any excess we find there will be extremely significant. As for the data we have now, it is not an *a priori* search, so any significance estimate is simply useless."

It was now Melissa Franklin's turn to express her views. She had arrived late to the meeting, and she had kept quiet, standing up next to the entrance door, with a few of her students sitting on the floor around her. She waved her hands as she expressed her views:

"You may be right — yet what I am wondering is why could we decide to keep for ourselves the information coming from the analysis of the Frascati group. Look at the big picture. We are more than 500 physicists deciding to hide some potentially interesting data, events which *might* be a signal of new physics, to the rest of the community of ... how many? 10,000 more particle physicists around the world? For what reason? I cannot understand it. Even if there *is* a mistake in the analysis, some obscure detector effect that nobody in this room is capable of even hypothesizing, why should we be scared to say what we see, and let others know? I think this is madness. Actually, I *hope* it is madness, because it smells of something even uglier."

The italics are words she stressed in her speech. I was totally with Melissa on that point. I hated that the collaboration would rather ditch a potentially important scientific result than run the risk of either being proven wrong, or being perceived as an experiment that indulged in the broadcasting of ill-supported claims.

The discussion on the superjet analysis continued for a while longer, mostly to allow Paolo to finish his talk — but the majority of the Top group members, which constituted a good part of the opinionated participants at the meeting, had by then made sure to show they were not convinced. Their influence in the arena was strong, and those who had been listening to the discussion without contributing to it were likely to accept their conclusions. It only remained for the spokespersons to wrap it up in a way which would not hurt feelings any further.

"I think we have heard enough for today on this matter". Said Goshaw. "Let me remind you that several months ago we appointed a special committee of six godparents, who have dissected the analysis of superjet events by contributing with real work, in a search for mistakes which have not been found. Asked on whether the analysis *should* be published they have been unable to reach unanimity, but they recommended that the collaboration *seriously consider* publishing those events. We are all very busy with the upgrade, and I see lots of energies being drawn away from our main goal, the commissioning of the CDF II detector. From what we heard today, I would summarize that there does not seem to be a consensus of our collaboration in either direction. In the absence of that, Franco and I believe we need an Executive Board vote to decide how to proceed. I would describe the options on the table as follows. Motion One is to second the authors' request and proceed with a paper draft of the superjet analysis, as with any other publication. Motion Two is to continue with the study of the superjet events, in a suitable way which is still to be determined, after consulting with the past godparents and the authors of the analysis."

"Just a moment, Al," Tony Liss interjected. "I would like it if we clarified a point here: regardless of the outcome of the vote, the analysis needs to be blessed before we can even start *thinking* of writing a draft. It has not passed that mandatory step yet, and we have not decided to change the procedure for this case, I don't think."

"Of course," replied Goshaw, "that is included in Motion One. I am not advocating any special treatment for this paper. If Motion One is passed, there will be a blessing procedure before a draft is circulated for publication."

"I would like to request that we also vote on a third Motion, namely, the one of dropping the examination of the analysis altogether," said Avi Yagil. "After all, that is not against what the godparent committee has asked us to decide: consider publishing, and decide one way or another."

Goshaw nipped the argument in the bud. "I do not think we can decide in that direction, Avi. The authors have the right to try and seek approval of their work. Our bylaws do not include the censorship of physics analyses. So let us see. Who is in favour of Motion One? Please, only institution representatives are allowed to vote. Raise your hands."

Only few participants raised their hand. Much too few — maybe six or seven. Harvard, Frascati, Pisa, Padova representatives among them.

"Ok." Goshaw did not even bother counting hands. "Far sites, anyone in favour of Motion One?" No answer. "Who votes for Motion Two then?"

At least twenty hands went up. A few more representatives pitched in from far sites connected via videoconference, expressing their preference.

"Abstained?" Some more hands went up. Irrelevant, but procedurally correct. "Fine, then. Motion Two is accepted. Franco and I will work out a reasonable plan to continue the review of the analysis. Those of you who have expressed concerns on specific issues today are invited to produce a written set of requests to the authors, who will then proceed to answer them."

The meeting was over, and its outcome was clear: the publication of the superjet events would be prevented for many more months. As a result of the discussion, and my inability to keep my mouth shut, I was now considered a partisan, by some even a "co-author." Avi, who was a specialist with colorful expressions, preferred to say that I had acted as a "lightning rod," shielding Paolo from some of the most pointed objections.

Epilogue

The reaction of the CDF institutions to the draft articles on the superjets and on the scalar quark, along with the heated meeting reported above and a few others that followed, clarified that a good part of the collaboration was not prejudicially against the Frascati analyses. What almost everybody claimed, however, was that the matter should be solved one way or the other: too much energy was being drained away from the preparatory activities for the new run of the Tevatron, which promised to deliver much more data at 2-TeV energy. A reason to hope for a resolution was the successful publication of the top-pair production cross section. That paper at least demonstrated that Frascati had done a good service to CDF on a very important measurement. Based on those positive factors, the Oversight godparent committee worked on the difficult task of converging to a solution of the superjet affair which would be accepted by the largest possible majority of the CDF members. The path of least resistance was to dismember the three analyses, by dividing them in smaller chunks. That would hopefully allow for an easier digestion by the collaboration.

First, the 13 superjet events were described in a paper meant for publication in *Physical Review D*, which only described the bare facts. The spokespersons bargained with Giromini that in exchange for stripping that publication of any new physics speculations, he would be allowed to publish by himself a separate article, where he could discuss his interpretation of the observed effects. This plan once again met fierce opposition by the few known arch-enemies of Paolo and was the source of more endless debates, but in the end it won over. The article titled "Study of the heavy flavor content of jets produced in association with W bosons in ppbar collisions at sqrt(*s*) = 1.8 TeV" was finally published in 2002, with the signature of all but a few collaborators.

A second paper, "Additional studies of the probability that the events with a superjet observed by CDF are consistent with the SM prediction," only signed by 15 collaborators, was published back-to-back with the first in the same journal. The latter was a short document only meant to assess the probability to observe such weird kinematics in the superjets sample. Not even that number had been allowed into the former article, so it found a place in the separate publication. One of the 15 names on the front page was mine: invited to sign the article by Paolo, I could not refuse. So Avi and the others were proved right at last — I had become a co-author!

After the publication of the superjet paper, the Godparent committees were dismissed, and CDF finally moved on. However, Giromini and collaborators continued to work to publish *all* their results. The paper which discussed the interpretation of the 13 events in terms of a new physics model was submitted in 2003, but then withstood three more years of iterations with the referees of the journal. The article was finally published with only the names of the Frascati group members. Meanwhile, a few months earlier, a fourth paper describing the inclusive lepton excess was published, again signed only by the Frascati group members. And finally, in 2005 the analysis of the dimuon mass distribution also made it to PRD, with the unassuming title of "Search for narrow resonances below the Upsilon mesons." The result described there was only an upper limit on bound states decaying to muon pairs; yet, the mention of the 7.2 GeV bump was there, along with the suggestive mass distribution shown in Figure 1 of Chapter 12.

With the delayed and piecemeal publication of the Frascati analyses, the controversy over the superjet events and all the related anomalies spotted in other datasets died out. Or so it seemed. As Dante Amidei had predicted, Run 2 would "be a maelstrom" in comparison with those amicable Run 1 skirmishes. But that is another story.

Acknowledgments

The story of the CDF experiment encompasses four decades as well as the lives of several thousand physicists, engineers and technicians who participated in the construction and operation of the detector and in the exploitation of data collected over a quarter of a century. Even though I have chosen to base this book mostly on the discussion of the anomalies that were unearthed in the analysis of Run 0 and Run 1 data, I have also tried to tell along the way the story of the experiment in the corresponding time range, including some account of facts that happened well before June 1992, when I first joined the collaboration.

Within such a large collaboration, too much goes on for a single individual to keep track of. It would, therefore, have been impossible to write these stories without the help of many friends and colleagues in whose company I spent 20 exciting years of my life. They provided important input and information that would have otherwise been lost. It is thanks to this "collective memory" that I have managed to reconstruct in detail many of the anecdotes reported in this book. In the hope of not forgetting any of the colleagues (both CDF members and other scientists) who have provided their help and recollections for the present text, I offer below their list in alphabetical order:

Dante Amidei, Mario Antonelli, Patrizia Azzi, Nicola Bacchetta, Stefano Belforte, Giorgio Bellettini, Sergio Bertolucci, Anwar Bhatti, Dario Bisello, James Bjorken, Robert Blair, Gino Bolla, Kevin Burkett, Claudio

Campagnari, Tiziano Camporesi, Marcela Carena, Bill Carithers, Andrea Castro, Pierluigi Catastini, Lucio Cerrito, Paolo Checchia, Giorgio Chiarelli, Frank Chlebana, Allan Clark, Marina Cobal, John Conway, Ray Culbertson, Luc Demortier, Paul Derwent, Simone Donati, Flavia Donno, Monica D'Onofrio, Umberto Dosselli, Steve Errede, Tom Ferbel, Brenna Flaugher, Melissa Franklin, Henry Frisch, Mary Gaillard, Lina Galtieri, Paola Giannetti, Nikos Giokaris, Paolo Giromini, Douglas Glenzinski, Hans Grassmann, Carl Haber, Eva Halkiadakis, Fabio Happacher, Robert Harris, Matthew Herndon, Chris Hill, Joe Incandela, Eric James, Hans Jensen, Tom Junk, Gordon Kane, Ben Kilminster, Young-Kee Kim, Jacobo Konigsberg, Eve Kovacs, Joseph Kroll, Mark Kruse, Steve Kuhlmann, Greg Landsberg, Sandra Leone, Joseph Lykken, Louis Lyons, Robyn Madrak, Petar Maksimovic, Michelangelo Mangano, Mosé Mariotti, Stefano Miscetti, Aseet Mukherjee, Jason Nielsen, Larry Nodulman, Riccardo Paoletti, Christoph Paus, John Peoples, Monica Pepe-Altarelli, Luisa Pescara, Marcello Piccolo, Jimmy Proudfoot, Giovanni Punzi, Ken Ragan, Pierre Ramond, Burton Richter, Michael Riordan, Luciano Ristori, Vadim Rusu, David Saltzberg, Andrea Sansoni, Michael Schmitt, Melvyn Shochet, Paris Sphicas, Luca Stanco, Roberto Tenchini, Paul Tipton, Slawek Tkaczyk, Dave Toback, Diego Tonelli, Mia Tosi, Marco Trovato, Nicola Turini, Barry Wicklund, Brig Williams, Brian Winer, Edward Witten, Steve Worm, Jeffery Wyss, Avi Yagil.

I also wish to offer special thanks to a few colleagues who must be singled out as they gave repeated and extensive input, shared documents, or offered advice. Here is a shorter list:

Steve Errede, Brenna Flaugher, Robert Harris, Ben Kilminster, John Peoples, Michael Schmitt.

Finally, I express my heartfelt thanks to Paul Ellingwood, Neil Bates, John Hasenkam, Gina Lee, Eleni Petrakou, Steve Schuler, and, in particular, Nisia Thornton for proof-reading parts of this book.

Name Index

Subject Index

A

accelerator division, 104
ALEPH collaboration, 259
all-hadronic decay, 71
Analysis_Control, 95
Anderson–Darling test, 202
antimatter, 3
antiquark, 2
Argonne National Laboratories, 35
assembly hall, 34

B

B0 building, 34
barn, 21
baryon, 4
batch queue, 73
BCDMS experiment, 213
beam constraint, 54
beam pipe, 30
beta decay, 8
big bang, 19
Bjorken x variable, 24
B hadron, 50
boson, 2

bottom quark, 4
b-quark tagging, 114
branching ratio, 79
bremsstrahlung, 33
Brookhaven laboratories, 3

C

calorimeter, 30
  electromagnetic calorimeter, 33
  hadronic calorimeter, 33
calorimeter crack, 236
CDFNEWS forum, 211
CDF note, 81
central muon system (CMU), 34
central muon upgrade (CMP), 34
central tracking chamber (CTC), 32
CERN laboratories, 14
CES detector (a.k.a. "shower max"), 114
charge clustering, 107
charged current, 7
charge exchange process, 162
charm quark, 4
Chernobyl disaster, 46
Chicago Tribune, 210
chi-squared statistic, 136

Printed in the United States
By Bookmasters